T0142928

Practical Networking

Series Editor

Zhi-Li Zhang
Sci & Eng, 4-192 EE/CS Bldg
University of Minnesota, Dept of Comp
Minneapolis, MN, USA

Explosive growth in cloud and mobile computing coupled with new advances in systems and networking technologies as well as machine learning and artificial intelligence (AI) have revolutionized how networks and distributed systems are designed, developed, operated and managed. This is epitomized by data center networking where it has spurred a wholesale rethinking and re-designs from network architectures, to physical interconnects to routing, flow management and network application support. New networking paradigms and technologies such as software-defined networking, network function virtualization, smart NICs and software/hardware co-designs have emerged for better designing, operating, managing and evolving networks, and also enabled new visions such as "self-driving networks" and AIOps. The Practical Networking Series is centered on emerging topics in new networking paradigms, architectural designs, algorithms and mechanisms for primarily wired networks (from data center networks, enterprise networks to ISP networks), but also touches on "packet core networks" for emerging 5G and beyond cellular and wireless networks. Books in this series address these topics from both theoretical (e.g., new theoretical foundations, algorithms and performance analysis) and practical (e.g., new network mechanisms, protocols, APIs and standards, software frameworks) perspectives. Relatively short books on a timely and focused topic, research monographs, and textbooks are of interest. The Editor is seeking well written works by well-established researchers and practitioners in the networking field around the world, particularly Asia and North America.

Prospective Authors or Editors:

If you have an idea for a book, we would welcome the opportunity to review your proposal. Should you wish to discuss any potential project further or receive specific information regarding our book proposal requirements, please contact Zhi-Li Zhang or Susan Evans:

Zhi-Li Zhang
Department of Computer Science Department
University of Minnesota
4-192 Keller Hall, 200 Union Street SE
Minneapolis, MN 55455-0159
zhzhang@cs.umn.edu
Susan Evans
Senior Editor
Springer Nature
233 Springe Street
New York, NY 10013 USA
susan.evans@springernature.com

More information about this series at http://www.springer.com/series/16325

Nitul Dutta • Hiren Kumar Deva Sarma
Rajendrasinh Jadeja • Krishna Delvadia
Gheorghita Ghinea

Information Centric Networks (ICN)

Architecture & Current Trends

Springer

Nitul Dutta
Computer Engineering Department
Marwadi University
Rajkot, Gujarat, India

Hiren Kumar Deva Sarma
Department of Information Technology
Sikkim Manipal Institute of Technology
Majitar, Sikkim, India

Rajendrasinh Jadeja
Electrical Engineering Department
Marwadi University
Rajkot, Gujarat, India

Krishna Delvadia
Information Technology Department
Chotubhai Gopalbhai Patel Institute
of Technology
Uka Tarsadia University
Surat, Gujarat, India

Gheorghita Ghinea
Computer Science Department
Brunel University
Uxbridge, UK

ISSN 2662-1703   ISSN 2662-1711 (electronic)
Practical Networking
ISBN 978-3-030-46738-8   ISBN 978-3-030-46736-4 (eBook)
https://doi.org/10.1007/978-3-030-46736-4

This Springer imprint is published by the registered company Springer Nature Switzerland AG
The registered company address is: Gewerbestrasse 11, 6330 Cham, Switzerland

Preface

Distribution and manipulation of information over the Internet has been one of the most popular applications of technology that society has witnessed in the last two decades. Again, the Internet is one of the most popular technologies that has been adopted by society in the last century. However, present form of the Internet suffers from several issues, mainly the scalability in terms of the number of users and the number of applications running over it. Therefore, a change in the communication model in terms of the content distribution technology has been adopted. As a result of this, peer-to-peer (P2P) and content distribution network (CDN) have emerged that promote a communication model in which data is accessed by name rather than referring to it by its origin, i.e., the server.

Increasing traffic volume, increasing number of mobile and video applications, cloud computing, social networks, and big data applications are major concerns for the current form of the Internet. It seems there is a need for evolution in the architecture of the Internet to cope up with this changing environment. Different CDN providers and P2P applications are based on the proprietary distribution technologies. Such distribution approaches are implemented as an overlay, and therefore, it is associated with different undesired inefficiencies.

Information Centric Networking (ICN) brings such an evolution to the current Internet architecture. The focus here is on named data objects. There is a shift of core principle in the present form of the Internet, i.e., **from IP address–based information source** to **uniquely named data** in ICN. Data in ICN becomes independent from location, storage, and application. In-network caching and replication of data is permitted in ICN. Issues like scalability and high-bandwidth demands of applications over the Internet are expected to be handled with efficiency in ICN. There are few implementations of ICN available in the form of TRIAD (Translating Relaying Internet Architecture Integrating Active Directories), NetInf (Network of Information), DONA (Data Oriented Networking Architecture), and NDN (Named Data Networking).

ICN is an evolution of the Internet and it is expected to get reflected in various layers of the protocol stack of the Internet. Name-based data access being the primary goal, ICN may evolve the Internet architecture in different layers.

Although ICN seems to be a fascinating idea, it has many research challenges. Following is a list of at least few such challenges that need attention of researchers. Novel schemes for naming of data, schemes for name resolution, routing protocols which are scalable, in-network caching mechanisms, congestion control mechanisms, quality of service provisioning, security and privacy provisioning, application level protocol design along with applications interfaces (APIs), and frameworks for business, legal, and regulatory purposes are few of the research challenges associated with ICN.

This book is an effort to put various stuffs related to ICN together systematically so that researchers and interested readers can get an integrated view of the subject. There are nine chapters incorporated in this book that cover most of the associated topics of ICN. The book starts with a chapter that sets the platform for ICN, comparing it with the Internet, and also reasons why ICN is now a necessity. Different available ICN architectures such as TRIAD, DONA, NetInf, and NDN are elaborated in the second chapter. The third chapter outlines various naming techniques adopted in ICN, followed by progress in routing protocols that have been developed for ICN in the last decade, in Chap. 4. Various caching techniques and the recent progress in this area are presented in Chap. 5. Security enhancement is described in Chap. 6. Optimization in ICN is covered in Chap. 7, and the integration of SDN into ICN is introduced in Chap. 8. In the last chapter, i.e., Chap. 9, few applications where ICN is extended are highlighted.

The book provides guidelines for the current research challenges and future trends in various areas associated with ICN so that researchers get ready reference. It is expected that researchers and readers will get adequate information on the subject, and the book will be helpful in their research endeavours.

Rajkot, Gujarat, India   Nitul Dutta
Majitar, Sikkim, India   Hiren Kumar Deva Sarma
Rajkot, Gujarat, India   Rajendrasinh Jadeja
Surat, Gujarat, India   Krishna Delvadia
Uxbridge, UK   Gheorghita Ghinea
14/Feb/2021

Contents

Chapter 1
Introduction to Information-Centric Networks

1.1 Introduction

Information-Centric Networking (ICN) is an effort to transform the current Internet architecture. The current architecture of Internet emphasizes on information exchange among machines apparently dependent on the naming system such as Uniform Resource Locator (URLs) and through Domain Name System (DNS) resolution. The goal of ICN is to shift the focus from connecting machines and fetching data from specific hosts or content producer, as in traditional Internet, toward fetching data from any location based on the content identifier or the content itself [1–3]. In this system, the data is named rather than naming a machine. During nineteenth century, services as well as innovations related to Information and Communication Technologies (ICT) have proposed different means for human and machine interaction. The basic and significant invention named "Internet" came into existence in order to provide effective communication across geographically dispersed users by connecting different computing resources on the globe [4]. In order to communicate across networks, various communication paradigms and associated protocols like Transmission Control Protocol and the Internet Protocol (TCP/IP) communication model [5] came as the solutions. In which, distinct nodes can interact with each other by establishing any communication paradigm between them. This approach provides an excellent solution for client server applications like File Transfer Protocol (FTP), Hypertext Transfer Protocol (HTTP), Simple Mail Transfer Protocol (SMTP), and telnet. The current Internet usage patterns, preferences, and demands show that the majority of Internet traffic is generated by videos due to evolution of video streaming portals like YouTube, Netflix, IPTV services, etc. Such bandwidth-hungry applications require a change in the current Internet paradigm from host-centric approach to content-centric approach. As, most of the time, similar contents are fetched by a group of people and in traditional networks all of them get the copy from the source. But in content-centric approach the user with a copy of the data may share it with her neighbors. It significantly reduces bandwidth wastage and

© Springer Nature Switzerland AG 2021
N. Dutta et al., *Information Centric Networks (ICN)*, Practical Networking,
https://doi.org/10.1007/978-3-030-46736-4_1

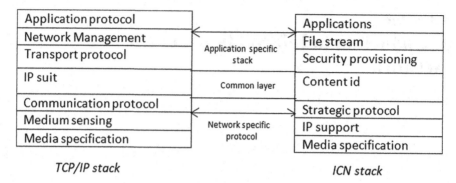

Fig. 1.1 Architectures of IP and ICN protocols

provides faster data delivery. For effective implementation of the new approach of Internet, there are changes made in the protocol stack of TCP/IP. Few protocols are modified and few new concepts are added to the so-called layered architecture of the Internet. A brief description of the changes is presented below.

Figure 1.1 provides a comparison of TCP/IP layers with Content-Centric Network (CCN) protocol stack [6]. Primarily, the thin waist of TCP/IP is replaced by the content name identifiers. Location-dependent node names are of no use in the new form of Internet. The fundamental principle behind data communication is client's name-based data request. It enables users to emphasis on the content that they demand, instead of referencing physical location of the content from where it can be retrieved. As data may be supplied by any node, hence security is a big issue. The data need to be self-secured. ICN protocol stack includes a security layer as an integral part of the architecture, whereas, in TCP/IP, security is included as an auxiliary protocol. Due to rapid growth in the usage of digital media, e-commerce, and social networking portal, the use of Internet is more of a distribution network, rather than communication network. To solve problems of distribution networks with the help of communication protocol, such as point to point, is more complex and causes errors. This simple conceptual change permits ICN networks to be applicable in order to solve broader range of challenges beyond end-to-end communication paradigm. The basic building blocks of ICN paradigm are named data. This is opposite to the IP network architecture's basic unit of communication, which is a point-to-point communication among two end endpoints recognized by IP addresses.

1.2 Internet Architecture: Strengths and Limitations

The Internet is network of networks. The present Internet architecture is a constantly and rapidly evolving interconnection of thousand networks. The Internet acts as a carrier that provides basic packet delivery services and without any guarantee. It means that Internet makes its **best effort** to try to deliver packets to the receivers that senders want to send. Internet Protocol (IP) addresses are used to identify end-hosts in the network.

The current Internet architecture along with its main supported functionalities are shown in Fig. 1.2.

Various nodes present in the Internet architecture are as mentioned below [7]:

(a) Content servers or caches (either professional or user-generated content and services)
(b) Centralized, decentralized, or clustered servers, including search engines and supporting servers
(c) Core and edge routers and residential gateways
(d) Users connected via fixed, wireless, or mobile terminals

The above-mentioned architecture works for current applications and usage. This architecture shall continue functioning and also be adequate if there are sufficient network resources. Here, network resources indicate bandwidth capacity, delay involved, latency involved, etc. Delay and latency involved in delivering packets are expected to be low. However, the same architecture as shown above would not be suitable if the number of connected devices is in the tune of billions and even more. Moreover, if users demand high-resolution videos, then that will require increased bandwidths. Again, if more and more users conduct delay critical real-time video and audio communications over the Internet, then the above-mentioned architecture may not be sufficient in order to provide that. Additionally, considering the massive growth in the number of mobile devices, if we shift into a wireless scenario the bottleneck increases seriously.

In present scenario, today's traffic is indeed driven by information production and retrieval through activities like simple RSS feed aggregators or advanced

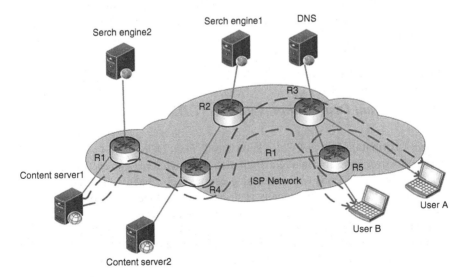

Fig. 1.2 Internet architecture [7]

multimedia streaming services, including user-generated or dynamic web content. The relationships between network entities are not restricted to the view of network topology, but also represent social or content-aware connections between users that can additionally share common interests (e.g., newsgroups, online photo, video sharing, and social networks). Such a situation makes the task of modeling today's Internet through graph a very complex one.

Strengths Internet is such a successful technology in present time that almost half of the population of the world today makes use of Internet. Considering the applications running over Internet, it can be concluded that almost everyone is influenced by the presence of Internet. Reaching everyone and present everywhere through Internet is the great strength of Internet. Internet is still allowing people to develop and run various applications over it for the benefit of the society. Internet is available through wired as well as wireless networks and people are deriving benefits out of that. These are some strengths of Internet which are very much evident.

Limitations In spite of various strengths of Internet, as mentioned above, the present form of it suffers from several severe limitations or constraints. As the applications over Internet are ever increasing, a situation shall come where current form of Internet will not be able to serve users satisfactorily. Quality of Service (QoS) shall be an issue. Even now Internet does not guarantee Quality of Services, rather it provides best effort. Current host-based architecture of Internet shall not be able to cater ever-increasing user base. A point of saturation is bound to come considering user applications and the user base, where services will suffer from high latency and reliability issues. Bottlenecks can occur even in the presence of content-based networks, since surrogate servers maintaining cached copies of various contents also come under host-centric paradigm. Current form of Internet suffers from various limitations, which are as follows: lack of data identity in the network, lack of proper storage management, lack of efficient transmission method of content-oriented traffic, insecurity in the transmission links, lack of efficient congestion control, lack of bandwidth in some network segments, etc.

1.2.1 Content Distribution Networks (CDN)

Content Distribution Networks, also known as Content Delivery Networks, evolved in order to overcome the inherent limitations of traditional Internet in terms of Quality of Services (QoS) while accessing various web contents. In CDN, a group of servers which are geographically distributed work together to provide faster delivery of web content. Thus, a contents in original servers are replicated and cached in several other servers which are scattered over the globe, with an objective to deliver content to the end users in time. To do so, nearby surrogates to the end users are used. Content distribution across the Internet has attracted considerable attention of the researchers. Such a topic involves high-end computing technology, high performance networking infrastructure, and efficient techniques for distributed replica management.

Content Distribution Networks boost network performance in terms of maximizing bandwidth utilization, improving accessibility, and also maintaining correctness of the contents through replication of it [8]. CDN can provide fast and reliable applications and services to the users. This is enabled through distribution of contents to the cache or edge servers which are located near to the end users. The content delivery infrastructure is consisting of a set of edge servers which are also known as "surrogates." These surrogates deliver copies of content to the end users.

Thus the limitations of current Internet architecture instigate evolution of it. Such an evolution is possible through the concept of Information-Centric Networks (ICN). Thus, ICN-based Internet will lead to much more efficient use of available network resources.

1.2.1.1 Content Distribution Network Architecture

A content distribution network has several components in its entire infrastructure. Those are as mentioned as follows: content delivery, request routing, distribution, and accounting. The content delivery infrastructure is a set of edge servers which are also known as surrogates. These servers deliver copies of the contents to the end users. The responsibility of directing client requests to appropriate surrogates lie with request routing infrastructure. The request-routing infrastructure interacts with distribution infrastructure in order to keep latest view of the stored contents in the caches of the CDN. Distribution infrastructure is responsible for moving contents from the origin server to various surrogate servers, and also ensures consistency of the contents in the caches. The accounting infrastructure is responsible for maintaining clients' access logs and also for recording usage of the servers in the network. This information is useful in traffic reporting and usage-based billing. Content distribution networks typically stores static contents that include images, video, media clips, advertisements, and also different embedded objects for dynamic web content. Generally, customers of a CDN are media and advertisement companies, data centers, Internet service providers (ISP), online music or video providers, mobile operators etc. These customers of CDN always want to publish and provide their contents to their customers or end users on time and reliably [8].

Following are the various services and functionalities a CDN would like to provide: storage and management of content, distribution of content among the surrogates, cache management, delivering static, dynamic and streaming contents, keeping backup and procedures for disaster recovery, performance measurement and monitoring, and necessary reporting. A CDN needs to build infrastructure for supporting all of the above services and functionalities.

Thus, CDN is a collection of various network elements aiming at effective delivery of contents to the end users. The entire ecosystem for a CDN may be homogeneous or heterogeneous. Moreover, CDN may be centralized or hierarchical under certain administrative control, or even entirely decentralized too.

To be more specific, various typical functionalities of a CDN may look like as mentioned below:

(a) Request redirection and content delivery services: This is to direct a request to a nearby surrogate avoiding congestion and bottleneck.
(b) Content outsourcing and distribution services: This is to replicate and cache content and also to distribute content to the surrogate servers.
(c) Content negotiation services: This is to meet the specific individual user's need.
(d) Management services: This is to manage the network components, to handle the accounting activities, and to monitor and report on usage of the contents.

Sometimes there is a sudden spike in the web content requests and this is called as flash crowd or SlashDot effect. A CDN can handle such a situation better in terms of network performance. This is achieved through caching of the web contents in some strategically placed surrogate servers in various locations. Thus, response time of the user requests are improved.

In CDN, content refers to digital data resources which are of two types. These are encoded media and metadata. Examples of encoded media are audio, video, document, image, webpages, etc. This type of content may be static, dynamic, or even continuous. Metadata is data about data. It involves content description which facilitates in identifying, interpreting, discovering, and also managing the data or content.

CDNs are actually a new virtual overlay to the Open System Interconnection (OSI) reference model. This layer can provide overlay network services. It relies on various application layer protocols such as Hypertext Transfer Protocol (HTTP) and Real-Time Streaming Protocol (RTSP) for data transport.

Figure 1.3 describes an abstract architecture of a CDN. There are three key components in a CDN architecture, namely, content provider, CDN provider, and end user. Content provider delegates the Uniform Resource Identifier (URI) name space of the web object or content which is to be distributed. The origin server of the content provider keeps those web objects. CDN provider is a company that offers infrastructure to content providers for delivering content in time and reliably. And finally, the end users access the contents from the content providers' websites.

Similarly, Fig. 1.4 is a schematic diagram showing various services offered by CDN [8].

1.2.1.2   Layered Architecture of CDN

The architecture of a content distribution network can be described through a layered model. Figure 1.5 is such a model composed of various layers that consist a CDN. Various layers present in a CDN are as mentioned below: Basic Fabric, Communication & Connectivity, CDN, and End User.

Basic Fabric It is the lowest layer in the layered architecture of CDN. This layer has several components such as Symmetric Multiprocessing (SMP), clusters, file

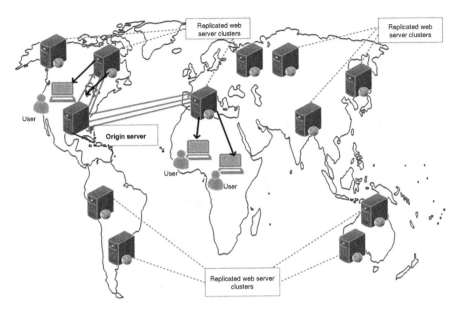

Fig. 1.3 Abstract architecture of CDN [8]

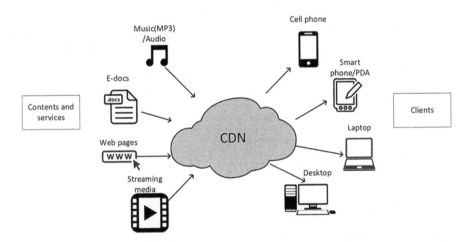

Fig. 1.4 Services offered by CDN [8]

servers, index servers, and basic network infrastructure. High bandwidth network is a necessity in CDN. These resources or servers run system software like operating systems, distributed file management systems, content indexing and management systems, etc.

Communication and Connectivity This layer maintains core Internet protocols, for example, TCP, UDP, FTP, and also protocols specific to CDN such as Internet

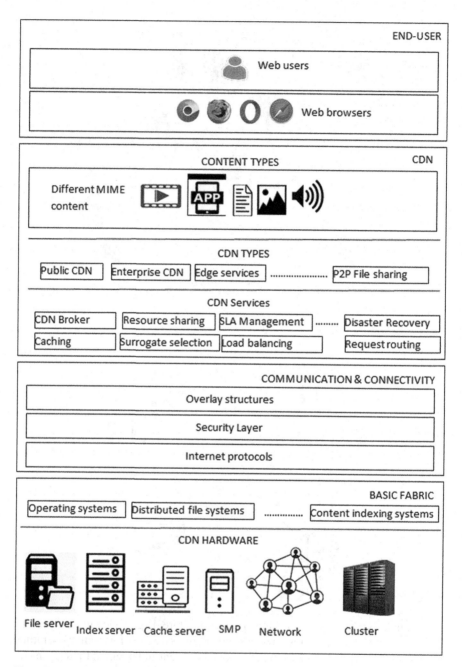

Fig. 1.5 Layered architecture of CDN [8]

Cache Protocol (ICP), Hypertext Caching Protocol (HTCP), Cache Array Routing Protocol (CARP), authentication protocol like Public Key Infrastructure (PKI), Secure Socket Layer (SSL), etc. This layer also ensures that delivery of content and services happen in an authenticated manner. Distributed indices are maintained to facilitate search and retrieval of replicated content.

CDN Layer Core functionalities of CDN are present in this layer. This layer can be divided further into three sublayers. Those are CDN services, CDN types, and content types. Various core services provided by CDN are surrogate selection, request routing, caching, geographic load balancing, user-specific services for Service-Level Agreement (SLA) management, resource sharing, and CDN brokering. A CDN may be a public CDN, enterprise CDN, edge server, or a server for file sharing based on a peer-to-peer (P2P) architecture. A CDN provides all types of contents such as text, image, audio, video, etc.

End Users Web users are the end users who connect to the CDN by providing URL of the website of the content provider.

1.3 Characteristics of ICN, Design Principles, and Assumptions

1.3.1 Characteristics of ICN

ICN has attracted attention of the researchers in recent times because of its numerous advantages and promises with respect to overall network performance. ICN strives to meet the demands of current Internet users. It also offers promises for the future Internet users. In this subsection, various characteristics of ICN are highlighted and discussed.

1.3.1.1 Content-Centric Communication

In contrast to traditional Internet Protocol (IP), ICN considers content as a first-class entity in entire network. Whenever an end user needs certain data from content source, it sends interest packet to the network, which contains name of the requested content embedded inside it. Based on this name, the packet is forwarded to node which contains the requested data. The node having the corresponding data packet sends back the content to the requester. When consumer sends interest packets to get data by their unique names, this interest packet does not contain addresses of source and destination inside it. Routers in the network need to forward those depending on the names of the content. It implies that each and everything in ICN is associated with content names and each individual action revolves around the content. Thus,

ICN supports content-centric communication rather than host-centric approach as it is in TCP/IP [5].

1.3.1.2   Location-Independent Networking

Before evolution of Internet of Things (IoT) [5], the research groups were trying their best to resolve issues imposed by existing TCP/IP paradigm of communication. The current Internet architecture has been modeled during the 1960s, where the purpose of communication was different and limited to communication between few connected resources across the geographically dispersed user base. But now the Internet usage patterns have changed drastically which include more bandwidth-sensitive applications. This actually demands shift into the new Internet paradigm called information-centric networks. Internet users are more interested in content rather than from where they are getting the content. The interest packet does not contain any location-dependent IP addresses associated to it regarding source or destination. The entire communication of interest and data packet takes place based on the name of requested data, rather than specification of explicit IP addresses. This makes ICN, a location-independent data-centric networking, because no matter where the desired content is located, without worrying about its location, it gets accessed through its name only.

1.3.1.3   Support for In-network Caching

In contrast to the existing Internet architecture that attaches IP address to individual node in the network, the ICN attaches unique name to content which is not relevant to its location. Request for data is forwarded to nodes having content, based on content name in place of IP addresses. This makes content-centric network approach more attractive to support in-network caching [9]. The in-network caching suggests each intermediate router in the network to cache a copy of the data that it had forwarded in the recent past. Whenever such a router having a copy of the content finds an interest packet for the same content, it sends the content from its cache, rather than forwarding the interest packet to the producer of the requested data. This reduces data delivery latency and traffic load on the network tremendously. It also increases the data availability. It reduces the amount of time required to retrieve the desired content after sending the request to the network. Moreover, multiple requests can be served by different nodes in the network having cached the same content. It further reduces load on one data producer to answer all the queries related to the same content. Hence, with provision of in-network caching, the problem related to content source bottleneck can be eliminated which in turns improves the overall network performance.

1.3.1.4   Name-Based Routing of Content

ICN is defined by isolating "where content is located" to "what actually content is." Because of this, routing in this approach demands to produce paths to different name prefixes that are associated with content, instead of address prefixes. In ICN, whenever routers forward interest messages, they build state information to send data packets back to the consumer. It constructs a reverse path from the state information maintained by router [10]. The router can explore distinct paths for individual name prefix based on the state information, maintained in the said router. Having facilitation of such multipath forwarding feature in the network system, ICN have made it easier to retrieve content efficiently and timely in a congested network. Moreover, in TCP/IP-based networks, for example, present form of Internet, the routing is based on IP addresses. On the contrary, routing in ICN is based on name prefixes in interest packets.

1.3.1.5   Cost-Efficient and Scalable Distribution of Content

The increased demand for extensive distribution and duplication of resources had made researchers to propose two prominent paradigms: the P2P approach and content distribution networks (CDN). In P2P, scalability in distribution is achieved by proposing self-organized, fault-tolerant, and adaptive distribution across different peers instead of giving entire load on servers. In content-centric distribution networks, the data is distributed on different caches administratively. So, the CDN approach is transparent to the end users as it redirects user requests for resources to caches. However, there are still some issues like network overhead associated to both the approaches. This leads to an emerging need for efficient and scalable content distribution. For example, if user is interested in fetching data disregard of its locations, it is possible to have architectural designs that address this requirement effectively [11]. Or there is a possibility to have paradigm that supports provision for in-network caching as well as content replica distribution across correct places at correct times. The proposed architectures for ICN so far effectively handles such requirements.

1.3.1.6   Unique and Persistent Naming of Content

As decisions regarding any action in ICN takes place based on the content name, it is important to decide the naming mechanism for data, which is unique, persistent, and independent of its location. Different naming schemes have been proposed based on flat and hierarchical namespaces (like URLs) with pros and cons of their own. The end user fetches content by its name; hence, it is very important to have unique name associated with each content chunk present in the network [12]. Moreover, authentication of objects via naming schemes is another requirement of ICN. This is so because the contents are not served from the original producer.

Rather it is supplied by any of the nodes that have the content in it. Therefore, integration of security provisioning in naming helps the architecture to use third-party security protocols as it is done in TCP/IP. The problem is existing authentication schemes like secure socket layer (SSL) and transport layer security (TLS) focus on authentication of endpoints of the connections in place of authentication of object itself. The prominent feature of content-centric networks is framed by keeping security parameter in consideration. Here the security constraints are applied on content instead of host; hence, it will be easier to validate the retrieved data. Because content is supposed to be self-validated with the help of public keys, there is no requirement for other security structures in order to validate publishers of content. Instead of it, user can show indication regarding trust on some specific publishers, and that trust is going to be spread to other data within a packet of data.

1.3.2   Design Principles

According to [7], it is mandatory to design ICN as a sustainable network, offering built-in support for energy-efficient solutions. ICN also needs to be flexible enough to continuously evolve over time. Similarly, ICN needs to be further developed and extended in response to the changing societal requirements. ICN should be scalable (i.e., it needs to serve very large number of entities), available (i.e., it needs to maintain its usable operation rate), and reliable (i.e., it should recover from faults if at all faults are there).

The major challenge in ICN is the task like distribution of billions of objects to billions of interconnected devices. It is necessary to have synchronization between contents in order to avoid naming conflict. The issue is to ensure globally unique naming to the contents. Therefore, name allocation to the contents is to be done through efficient procedures which are distributed and self-manageable and can handle huge number of dynamically generated content objects. Another design principle is that network management should be simple and better in terms of efficiency. Self-configured and self-optimized networking is the goal to achieve.

ICN is expected to be loosely coupled. Interdependencies between components should be minimum in spite of the large size of the network. Loosely coupled systems are flexible than tightly coupled systems. Users of Internet are more interested in having access to the desired data than the location of data (i.e., where the data is stored), Uniform Resource Locator (URL), or the IP address of the server that hosts the data. Thus, focus should always be on data or content. This is the basic design principle of ICN paradigm. Content advertisement, content discovery, and content retrieval should be easy in the ICN. While implementing ICN, content-centric routing needs to be supported. Content-centric routing protocol shall facilitate fetching of particular content to an end user, forming the most convenient location around the end user. Objective shall be to minimize the latency involved in the matter.

Achieving Quality of Services (QoS) in ICN is a challenging objective. QoS is also associated with the ability to use the available resources optimally. Since the notion of end-to-end flow does not exist in ICN, QoS needs to be defined with respect to objects or contents. This raises the complexity involved in designing QoS-aware ICN. This is so because contents are pervasive, replicated, cached, and distributed. Moreover, the contents can be accessed or even originated from many sources. Thus, contents can be brought via different paths. Such a situation leads to a challenging problem when QoS is under consideration.

ICN involves in-network caching. Multiple parties like content and service providers, ISPs, public and private organizations, regulators, end users all take part in an open environment. The concept universal and in-network caching is the prime mover of ICN. Thus, it is felt that caching should exhibit the following principles. Caching should be uniform (i.e., it applies to content carried by any protocol), democratic (i.e., not just for contracted Content Providers), and pervasive (i.e., all network nodes). Caching is the cost-saver form different perspectives like latency, retrieval time, etc. Finally, security needs to be ensured in ICN. Since end-to-end communication channel is out of consideration in ICN and, therefore, traditional security layers cannot be imposed in ICN in search of security in the network. Thus, security has to be integrated into the framework of ICN and content-level security needs to be achieved.

1.3.3 Assumptions

ICN was devised for wired Internet initially. This is the reason why various methodologies, protocols, and tools adopted by different ICNs are not very much suitable for wireless environment. Thus, there is a need to improve these for making suitable for next-generation wireless scenario. Most of the ICN implementations assume that operations of the network will be in a fixed network topology. The caching nodes are assumed to be resource-rich. Contents are also assumed to be static and long-lived. Contents were expected to be preserved as long as possible. Such assumptions may work well within wired Internet communications, in present scenario, contents are dynamic and their lifetimes are dependent on human activities in the environment. Moreover, scope of different content may also be limited to some geographic locations. Thus, a lifetime of contents is variable. In modern scenario, all caching nodes may not be resource rich, if at all few such nodes present in the network are aimed at exploiting strategically for the purpose of ICN. Thus, newer implementations of ICN should focus on finding best caching nodes for respective contents considering the lifetimes of the contents and strategic locations of the nodes.

1.4   Key Building Blocks of ICN and Design Challenges

The ICN is going to change the access patterns of bandwidth-sensitive Internet usage for the users. This new paradigm has opened the doors for the researchers to propose effective protocols and procedures for various issues including naming, routing, and security mechanisms. Due to the numerous technical needs in these protocols and procedures, design of such protocols and procedures are not straightforward. In this subsection, some of the challenges in designing ICN are enlisted along with brief descriptions.

1.4.1   Key Building Block of ICN

Data objects or contents are the building blocks of ICN. Everything in ICN revolves around data or content. However, there are few tasks associated to these data or content which can be considered as building blocks for constructing an ICN. Those are as mentioned below: naming of contents, content storage, routing, caching, content forwarding, and ensuring security in the network.

1.4.2   Design Challenges

Various challenges to be faced during the design of ICN are discussed in this subsection.

1.4.2.1   Content Naming

Naming of objects in ICN is a prominent feature which increases data availability of data and decreases delay to retrieve it. Let us consider a situation where user is requesting content by its name through interest packet. This interest packet is sent to content source. The content source replies to the interest packet with corresponding data along with the content name. Taking into consideration the in-network caching strategy, the communicated data may get cached at all intermediate locations along the reverse path from content source to requester. So, in future, if same or other node requests the same content, it can get satisfied by that nearest intermediate node in-between. But issue here is to provide a unique name to each of the contents available on the Internet. However, it leads to a certain situation that needs special care to implement ICN [10, 13]. For example, names assigned to contents must be globally unique. The name also has to identify the producer of the content for the ease of finding and authenticating the producer. Moreover, existing routing mechanisms are IP based. There is a need to have name-based routing mechanisms to support ICN. A set of strong, efficient, scalable, and secure naming mechanism for content must be proposed and implemented for efficient deployment of ICN. Naming should be in such a way that the size of the routing information table in the router is manageable and obeys some hierarchical structure.

1.4.2.2   Caching of Content

Caching of contents at intermediate nodes in ICN increases efficiency. But the challenge is how to accomplish this task effectively. There is always a choice of either caching every chunk of data or select some of them. Moreover, either data may be cached in all the intermediate routers or in some selected locations. All such options have their merit and demerits. Furthermore, cache size provision for nodes in network caching is another issue. The current emergence of video portals in Internet traffic leads to huge amount of data production at the end. In such cases the storage space in a router is always a constraint. These all should be taken into consideration in ICN implementation [11, 14]. It is a big challenge to propose an effective caching strategy for ICN architecture which can handle all such issues and provide an optimum solution for the same. A set of rigorous experiments on ICN tested with respect to all different input conditions is highly essential. The following are the few challenges associated with the content caching in ICN.

Placement of cache: Determining the location of the cache to store content is a challenging task. Caching may be either in-network or at edge in nature. No matter, either in-network or edge, both have certain limitations and hence determining optimal location of the cache storage is a challenging task. If it is edge caching strategy, then for content retrieval, there is a need of redirection of requested packets traversing a long distance to the requester. In such caches, location should be decided by certain strategies, so that there is less delay due to redirection of requests in order to fetch cached data. On the other hand, in-network caching strategy is simple to deploy but the decision regarding selection of nodes that should be equipped with fixed size of cache is an issue, because it is dependent on many other constraints related to topology and characteristics related to network traffic. These are open issues for both of the caching strategies. The cost also considered as a major factor during deployment, in order to operate the cache memory with line speed. There is a demand of faster memory which indirectly increases cost related to implementation. Based on decided capital investment, Internet service provider may decide cache placement strategy by considering other factors, like node centrality, expensive interdomain communication links, traffic patterns, domain size, etc.

Content distribution: Content distribution states the pattern in which a particular content is stored in the network or in the edge. Such distribution of content influences various parameters related to system, like redirection of request packets and system gain in terms of cache hits. The straightforward way for placement is to cache every content in each of the node in the path. It decreases the communication delay as well as message overhead. But it may lead to filling the caches completely. There will be a need of replacement of the cache content as the content is too large compared to the available capacity of the cache storage. With this increase of copies for a particular content inside the system, there is also a need to effectively manage the resources in network, which in turn is an issue [9]. Moreover, copying every piece of data in every network node leads to a redundant replication of data which is undesirable. There is some coordination among network elements needed before they take decision regarding caching content. So, reducing redundancy in content in the cache is also a challenge. Another issue is the nature of the data to be cached. Popular contents are required to be stored more frequently as compared to less popular data. During formulation of network caching strategies, expected content popularity should be taken into consideration.

1.4.2.3 Data Integrity and Origin Authentication

To name a data object in ICN is as essential as assigning location-dependent IP addresses in existing Internet architecture. Since these data objects can get cached at any location, the validity of data cannot be considered obvious. It is important to create verifiability between object name and the object itself which is termed as *data-name binding verification*. It is a fact that the intermediate sender other than the producer can make sure whatever the received data in recent past is the same named data object that is initially requested. This is called data integrity. The authentication for the origin of content is a distinct security mechanism that is associated with content naming. This means that to verify the content in reality, a named data object is published by the authorized respective publisher only. This may be recognized by associated name prefix. It needs to be assured for ICN in order to work reliably [15]. If this fails, then no one will trust the authenticity of data object, neither users nor the network entity. The lack of trust factor may lead to occurrence of various attacks like denial of services (DoS). There are distinct approaches to use cryptography with object names and manage the namespace for the object, accordingly.

There are two kinds of naming methods proposed in literature: flat naming and hierarchical naming. In the flat approach for naming, consistent names are assigned to content without maintaining distinct hierarchy of producer's identity. This scheme needs public key concept to verify integrity of object's name information. It demands inclusion of the publisher's public key in data object name and place signature for hash of the content along with associated private key. The resultant names are flat, though the field of publisher can be configured to produce a structure that can provide aggregation of route. Hierarchical names are identical to existing URIs, rooted with prefix of publisher of content. This scheme enables data aggregation in routing. Scalability of routing approach is also increased in this approach. In some situations, names are in human readable form that can be formed by end users manually. Such names can be reused and can map the object name to intent of end user. There are various trade-offs of designs in ICN that can influence security and data routing in ICN. Few of the naming specific challenges are explained next.

Naming dynamic content: Naming data objects is accomplished by content hashing as subset of name of the object. The publisher computes hash value on existing objects. An ICN node can verify the binding of content name by recomputing the hash value and compare the respective name. The name of the object has to be produced before the data object gets produced. In case of dynamic data like live streaming, it is quite difficult to provide name to such data [10]. In such cases, the publisher desires to make the data stream available immediately after generation. The naming of data object through registration of chunk of stream names in the network is quite time consuming.

Privacy preservation: The protection for requestor's privacy is also challenging in information-centric network because of name–data–objects–access approach. In general, the name of content should not disclose any information related to data. But in ICN the name is used to fetch the related objects of data. Moreover, the network is able to visualize the responses and requests through various logs maintained related to the history of requests from end users, which is not desirable.

Updating named data: ICN suggests not to change data names to avoid inconsistency issue. Because, after a copy of named data is created it is stored in the network for later retrieval. It demands the binding of content name to be consistent and should not get changed. It is not possible to change the content without creating related new name (updating). Updating may be done by versioning which is not supported by current naming.

1.4.2.4 Scalability of Resolution System and Name-Based Routing

The routing in ICN means finding routes to node which has named data objects based on content name provided by sender. The routing process is comprised of three different processes named as *resolution of name*, *route discovery*, and *content delivery*. In first phase, requested object name is translated into corresponding locator. In second phase, routing of data object based on its name takes place. In last phase, the data object is routed back to the requester (the client). The routing methods in ICN is categorized into lookup by name routing (LBNR), route by name routing (RBNR), and hybrid routing (HR). The challenges in all the three categories are described next.

Challenges in LBNR: This routing method uses first phase of name resolution to convert requested data object name into corresponding locator (IP addresses). The second phase, that is, discovery phase is based on existing IP paradigm, as route discovery is performed depending on locator. The third phase of content delivery can be implemented in the same way as that is done in IP-based routing. The locator of the sender is added in the request packet and requested content is sent back to sender depending on locator. The following are the challenges in LBNR.

Challenges occur in building the scalable resolution system during lookup and update. The lookup process performs mapping of data objects to its corresponding locators and make copy of it. Designing an optimized solution in this case is a considerable issue. Moreover, data objects location may get changed very frequently.

More than one data objects alter their locations at same instance of time. Getting adjusted with such changes is complicated. Creating in network replica during routing is another challenge. Caching as well as replacement of caches need to be done optimally to avoid frequent cache miss and efficient utilization of router storage.

Challenges in RBNR: This routing method excludes the first phase of name resolution and route a request based on the name of data object toward the content itself. So, routing table should contain information of routing for individual object of data. The size of a routing table is an issue as there is huge amount of data objects unless any mechanism related to aggregation comes into picture. This routing method reduces the overall delay and makes the routing process simple due to exclusion of name resolution phase. For third phase, this method requires other ID of any location or host to send the requested content back to the sender. Or else an extra routing method needs to be introduced. The following are the challenges for RBNR:

In order to decrease the entries in routing table, aggregation of names for data objects is difficult and needs special attention. Knowing about the name that is framed for the aggregation by content provider is another challenging task in RBNR. For example, let's assume that the name given to a content is as "ICN-challenges," and the content provider is ABC. At a later stage it is decided to alter the name to another name, for example, "/IETF/ABC/ICN/challenges," for the aggregation. In that scenario, intimating users about this change is difficult without a resolution method.

Challenges in HR: This routing method combines previous two methods to include benefits of from both. For the single-network domain, let us consider an Internet service provider (ISP), where the issues regarding scalability can be resolved by planning of the network. To decrease overall delay, RBNR can be used to exclude name resolution phase. In order to route the packets between different domains that contains their own locators, LBNR will be used. The following are the challenges particularly related to HR.

- Framing a scalable mapping system by giving the named data object and subsequent return of locator are complicated tasks. A destination domain should request content by encapsulation.
- Assuring secure mapping of information in order to prevent any malicious node from doing hijacking of the request packet is a challenging task.
- In case of content name alteration, verifying origin of data and ensuring data integrity are problematic.

1.5 Benefits of Using ICN

ICN node finds content by the name of the content and a node needs not to have any knowledge about the location of the content. It uses a managerial packet called "interest" to find the required content. The network routes this request in the entire network. Any intermediate router having the requested content generates the corresponding data packet and sends back to the requester. If no intermediate router holds the data, then the client is served by the source or the producer of the data. The two fundamental advantages of ICN over host-oriented approach are as follows: reduction in network load/traffic and reduction in delay involved [5]. As ICN supports caching of content at intermediate nodes in the network, the request for the same content by different clients can be served by different nodes in the network that have already cached the content previously, or by the content source itself. These activities actually reduce the network traffic by diverting it to multiple possible nodes in the network instead of single content source. Because of caching support, the delay involved in content retrieval is less as there is no or minimum queue of pending interests at single content source. It also reduces content retrieval delay and ensures high content availability due to provision for content caching in the network. There are several benefits of ICN-based architecture as compared to that of TCP/IP-based network architecture [16]. Few of such benefits are enlisted below.

1.5.1 Content Identity and Location Separation

Existing Internet is based on the identity of the participating nodes recognized by its IP address. Moreover, IP address is geographical location dependent, or subnet dependent. Hence, when a source of data is moved from one place to another geographical location, then the source of data get changed. As a consequence, data source needs to be registered newly or some mobility-related protocols need to be used. Due to this, it becomes difficult to reach systems that are mobile. Although, various solutions to this problem have been proposed in the past, which include node-identity protocol, mobile IP, etc. [4], still many subsequent problems of seamless mobility exist. The advent of ICN may effectively address these issues by providing requested data from intermediate cached locations rather than serving from the actual source.

1.5.2 Energy-Efficient Interaction

The communication paradigm requires to have active status before initiating a data transfer. In case, the receiver is in sleep mode, it must be awakened up to accomplish the communication. Otherwise, all the received data packets will be dropped. Although in the evolution of wireless nodes, this limitation disappeared by permitting base stations to cache the data packets while the requester node is in sleep mode [3]. It still exists in wired network. With the progress of ICN, and in-network caching of data, the said problem is possible to resolve up to a great extent. In this case, every intermediate router capable to do caching will act as a base station in wireless network, so far caching is concerned. Thus, energy of sleeping nodes in the network may be saved without losing any data.

1.5.3 Person-to-Person Communication

Although communication happens between computing devices, still the actual reason behind existence of communication networks is to have the same between human beings. A person is reachable through a wired computer, mobile phone, or laptop computer. The primary objective is to reach out that person, not the computers, phones, or the laptops. But in order to provide communication facility, both endpoints should have IP addresses. So, the person must select a device for person-to-person interaction in place of the actual destination node, which means the person. Since the ICN emphasizes on content identification and concept of IP address is avoided, hence it will enable to identify a person based on the used content. In that case, person-to-person communication will be easier because the person may be identified by his or her devices used, where earlier contents were delivered [17]. Moreover, the security provisioning in ICN ensures the authenticity of a person in the communication.

1.5.4 Explicit Help for Distributed Services and Traffic Across Client–Server

A huge amount of today's Internet traffic is due to client—server traffic. One example of distributed service model is as mentioned below: One user is trying to access Google, which in turn is not a single system [2]. In reality, it is a type of distributed services that have multiple systems located at multiple different places. End users are directed to nearest service providing servers for faster responses. However, because of associated location with a server, all users in the locality will be directed to the same server and hence distributed concept is not used in real sense. But if content is delivered based on the content identification then the same data may be served from different locations, hereby realizing the distributed delivery of content or service.

1.5.5 Integration of New Framework

In today's Internet, control plane, management plane, and data plane are merged together. TCP connection establishment packets are control packets and simple network management protocol (SNMP) packets are management packets. All these packets follow the same communication path that is followed by actual data packet [6]. In addition, control packets are also piggybacked on data packets in most of the cases. Because of this, there is much vulnerability and security threats as demonstrated by various attacks related to security in Internet. On the contrary, the network of telephone uses different networks for control packets, because of which it is considered secured than existing Internet. In order to get solution of this trouble, the data and control planes are separated by generalized multiprotocol label switching (GMPLS). This separation leads to the advantage of permitting data plane to become non-packet-centric, like the frames of SONET, wavelengths, and lines of power transmissions. This isolation is expected to be integrated in the future generation of Internet.

1.5.6 Privacy and Authenticity

Privacy preservation and authenticity are two major issues with the current Internet technology. It is provided to data communications through auxiliary protocols. Many researchers have proposed significant and prominent solutions to address such issues, but the issue is not resolved up to the full extent. The upcoming ICN technology have come up with the solutions of self-authentication of contents [17]. It will certainly address the above stated two issues. The need of third-party protocols for security provisioning would be omitted in this new technology. Moreover, due to the use of public and private key cryptography, privacy and non-repudiation may be addressed satisfactorily.

1.5.7 Asymmetric and Symmetric Protocols of Internet

In the current Internet architecture, most of the protocols are symmetric in nature [18]. It is framed for end systems with same configurations. In reality Internet is the connection of many heterogeneous devices, including handheld or palm devices to wireless sensors. In such scenarios, end systems are a constrained in terms of resources. So, it is reasonable to permit use of asymmetric protocols in future generation of Internet. Therefore, ICN is expected to support such sophisticated asymmetric protocols.

1.5.8 Requirements for Isolation

The end users demand isolation for the application in shared scenario. It is true, specifically, in applications that are critical in nature like distinct monitoring and military applications. Here, isolation implies that the performance of any one application should not get affected because of any other applications that share the identical resources in between. Practically, it is difficult to achieve [5]. Still, one alternative solution is to allocate dedicated resources to this kind of applications. To make this happen, virtual private networks are promising choices. In continuation with this, the future Internet based on ICN is aimed at providing programmable mixtures of sharing services for the application and subsequent isolation to the end users.

1.5.9   Quality of Service (QoS)

The QoS is defined as the extent to which an end user is satisfied with a service. As the name suggests, it is related to a service that is directly or indirectly associated with data that affects the provided services. Needless to say, objective of any domain of communication is to give satisfactory services to the end users through wired or wireless architectures [3]. Hence, guaranteed throughput and delay of packets are the prime concerns during an ongoing sessions between any pair of source destination. The guarantee for QoS is difficult to achieve in traditional IP-based networks. In contrast, the future Internet architecture has provision to permit guaranteed QoS parameters, before establishment of the sessions.

1.5.10   ICN and Internet of Things (IoT)

Internet of Things (IoT) is an interconnection among tiny battery-driven devices that can compute, sense, and communicate, and attached to an object that actually connect them with the world of Internet. These tiny devices are nothing but sensors that are smart enough to transfer and get data to or from another device. Instead of traditional point-to-point communication approach, IoT mainly focuses on information and data. Smart applications in IoT need contextual data that is produced by these sensors, reactively or proactively. Internet of things attracts plenty of research interests because of its wide domain of applicability. In the context of IoT, future Internet paradigm called ICN is also investigated and the various possible solutions are suggested by different researchers. Few of such works are presented here. This finally shows that ICN is beneficial even for IoT applications to emerge in the future.

In [6] a detailed discussion of ICN-based IoT is presented. The content of IoT is assigned unique name and it utilizes the benefits of ICN approach for IoT-based smart automation. A push-oriented communication is projected with the smart home automation concept. Another work that implemented IoT in the context of ICN is discussed in [15]. They state various constraints of IoT devices related to power, memory, processing, and sensing as hindrances to implement ICN-based approach in IoT. So, some functions like caching and security have been delegated to third-party trusted entity.

The authors in [9] presented a proposal for IoT-based content-centric network. Authors have used named data networking approach for an IoT-based environment with more than one content sources. The approach is named as Single Interest Multiple Data (SIMD). The authors have contributed a collision avoidance method by utilizing the contention window concept. Here, a new interest message is defined that requests content from various content sources, which is named as multisource interest (msINT). Authors have simulated the proposed approach and calculated various parameters such as interest packet overhead, time required for data collection, and number of interest packets. They also compared the performance of proposed approach against single interest and single data (SISD) scenario.

There have been numerous applications related to Internet of Things and have been visualized for broader set of applications like intelligent healthcare system, intelligent transportation and retailing of various goods, smart transportation, sensing of data that is cooperative in nature, interconnection among infrastructure and buildings, as well as monitoring/tracking of the same, smart manufacturing processes and smart products, smart home, smart sources of power and smart grid, intelligent interconnected city, etc., to name a few [19]. Many of the difficulties faced by users in traditional Internet will be addressed by new ICN technology. More importantly, the prominent benefits will be the reduction in network load and the delay experienced by end users while receiving the desired contents.

1.6 Historical Note on ICN

The 17 rules proposed in the Ted Nelson's Project in 1979 [20, 21] is marked as the first proposal of ICN in literatures. In the same year, the Named Data Networking (NDN) was initiated by the same author. The NDND is later carried out by Brent Baccala [17] and submitted as Internet Draft during 2002. However, in between, the TRIAD project [22] was initiated in 1999, at Stanford University. During 2006, the Data-Oriented Network Architecture (DONA) [23] project was initiated at UC Berkeley. The ICSI came up with content-centric networking approach, which is actually an improvement over the TRIAD approach. The new proposed architecture adds persistence and authenticity as first-class feature in the paradigm. In the same year of proposal of DONA, that is 2006, Van Jacobson [17] presented an architecture for NDN as the next step toward the growth of ICN. During 2009, Jacobson, research scholar at PARC announced their content-centric paradigm within the project CCNx [24]. PARC published interoperability specifications and gave open-source implementation of CCN project during September 2009. By then, there were various paradigms based on ICN which have emerged. The NDN paradigm is just one instance of it. During 2012, the research working group for ICN was established by Internet Research Task Force (IRTF). NDN has 16 major investigators which are funded by National Science Foundation (NSF). It is spread over 12 different campuses and gaining research interests from the industry as well as academic institutions. Global NDN test bed has been formed by more than 30 institutes across the globe. Detailed discussions on these topics are found in Chap. 2.

References

1. Handley, M.: Why the internet only just works. *BT Technol. J.* **24**(3), 119–129 (2006 July)
2. Stuckmann, P., Zimmermann, R.: European research on future Internet design. *IEEE Wirel. Commun.* **16**(5), 14–22 (2009 October)
3. Rexford, J., Dovrolis, C.: Future Internet architecture: clean-slate versus evolutionary research. *Commun. ACM.* **53**(9), 36–40 (2010 September)
4. Pan, J., Paul, S., Jain, R.: A survey of the research on future Internet architectures. *IEEE Commun. Mag.* **49**(7), 26–36 (2011 July)
5. Nagle, J.: RFC-896 "Congestion Control in IP/TCP Internetworks", ARPANET Working Group Requests for Comment. DDN Network Information Center, SRI International, Menlo Park (1984)
6. Waltari, O.K.: Content-Centric Networking in the Internet of Things. MSc thesis, Department of Computer Science, University of Helsinki, 25 Nov 2013 (2013). http://hdl.handle.net/10138/42303
7. Almeida, F., Loureno, J.: Information centric networks-design issues principles and approaches. *Int. J. Latest Trends Comput.* **3**(3), 58–66 (2012 September)
8. Pathan, A., Buyya, R.: A taxonomy and survey of content delivery networks Technical Report. University of Melbourne (2007)
9. Chai, W.K., Katsaros, K.V., Strobbe, M., Romano, P., Ge, C., Develder, C., Pavlou, G., Wang, N.: Enabling smart grid applications with ICN. In: 2nd ACM Conference on Information-Centric Networking (ICN 2015), 30 Sept–2 Oct 2015, pp. 207–208 (2015)
10. Varvello, M., Schurgot, M., Esteban, J., Greenwald, L., Guo, Y., Smith, M., Stott, D., Wang, L.: SCALE: a content-centric MANET. In: 2013 IEEE Conference on Computer Communications Workshops (INFOCOM WKSHPS), 14–19 April 2013, pp. 29–30 (2013)
11. Yu, K., Arifuzzaman, M., Wen, Z., Zhang, D., Sato, T.: A key management scheme for secure communications of information centric advanced metering infrastructure in smart grid. In: IEEE Transactions on Instrumentation and Measurement, vol 64, no 8, Aug 2015, pp. 2072–2085 (2015)
12. Amadeo, M., Molinaro, A.: CHANET: a content-centric architecture for IEEE 802.11 MANETs. In: 2011 International Conference on the Network of the Future (NOF), 28–30 Nov 2011, pp. 122–127 (2011)
13. Ren, Z., Hail, M.A., Hellbruck, H.: CCN-WSN—A lightweight, flexible Content-Centric Networking protocol for wireless sensor networks. In: 2013 IEEE Eighth International Conference on Intelligent Sensors, Sensor Networks and Information Processing, 2–5 April 2013, pp. 123–128 (2013)
14. Yu, K., Zhu, L., Wen, Z., Mohammad, A., Zhou, Z., Sato, T.: CCN-AMI: performance evaluation of content-centric networking approach for advanced metering infrastructure in smart grid. In: 2014 IEEE International Workshop on Applied Measurements for Power Systems Proceedings (AMPS), 24–26 Sept 2014, pp. 1–6 (2014)
15. Katsaros, K., Chai, W., Wang, N., Pavlou, G., Bontius, H., Paolone, M.: Information-centric networking for machine-to-machine data delivery: a case study in smart grid applications. IEEE Networks, vol. 28(3), pp. 58–64 (2014)
16. Ahmed, S.H., Bouk, S.H., Kim, D.: Content-Centric Networks an Overview, Applications and Research Challenges (2016) ISBN: 978-981-10-0064-5
17. Jacobson, V..: http://yuba.stanford.edu/cleanslate/jacobson.pdf (2006 February)
18. Named data networking, From Wikipedia, the free encyclopedia. Online available https://en.wikipedia.org/wiki/Named_data_networking
19. Wilson, S.: Rising Tide—Exploring Pathways to Growth in the Mobile Semiconductor Industry, 6 Nov 2013 (2013). http://dupress.com/articles/rising-tide-exploring-pathways-togrowth-in-the-mobile-semiconductor-industry/
20. From Wikipedia, the free encyclopedia online available https://en.wikipedia.org/wiki/Content_centric_networking

21. Director's Cut: Ted Nelson on Hypertext, Douglas Englebart, Xanadu and More, IEEE Spectrum, February 14, 2018
22. Gritter, M., Cheriton, D.R.: TRIAD: A New Next-Generation Internet Architecture (2000 July). http://www-dsg.stanford.edu/triad
23. Koponen, T. et al.: A data-oriented (and beyond) network architecture. In: Proceedings of the 2007 Conference on Applications, Technologies, Architectures, and Protocols for Computer Communications, August 27–31, 2007, Kyoto, Japan (2007)
24. Saadallah, B., Lahmadi, A., Festor, O.: CCNx for Contiki: Implementation Details Technical report RT-0432, p. 52. INRIA, Paris (2012)

Chapter 2
Architectures for Content Communication

2.1 Introduction

A system architecture splits the technological concept into functional units. The specification of such functional units and the interfaces among them enable researchers and developers to understand the working principle [1, 2]. It supports realizing the objectives and subsequent hinderances of the technology for successful deployment. It also helps in identifying interrelation among various structural components of the model to overcome common issues. However, the emerging functional architecture of a model must be examined and validated to ensure satisfactory feasibility of the proposed model from software or protocol perspective. These facts are equally applicable to ICN concept. ICN suggests unique naming of every piece of content with an intention to make it publicly reachable or searchable with the given name, globally. It treats data or content as the class one object for the deployment of the architecture. Hence, to understand the feasibility of the model, an effective architecture is highly essential. In the last decade, ICN has undergone a significant change in its architecture and several models have been proposed. All of these approaches differ in terms of implementations, but they have the same goal of improving performance and end-user experience by accessing the contents and services by name rather than by original location. In reality, the concept of data-centric network is not new. This concept actually comes from the concept of deploying mirror sites for web services where popular websites handle crowded clients by redirecting requests to its mirror sites. It helps in balancing load on servers and network as well to resolve overwhelming of resources in the server. In this case, the requested data is not necessary to be found in actual location; rather it could be taken from the location, where it is available. Here itself the content routing in ICN comes into existence.

Content-centric communication has gone through a tremendous resurgence in recent times. ICN offers advantages to end users and service providers for various dimensions like security, mobility, and performance. ICN achieves these advantages by bearing nontrivial cost because many ICN projects add significant complexity

© Springer Nature Switzerland AG 2021

N. Dutta et al., *Information Centric Networks (ICN)*, Practical Networking,

https://doi.org/10.1007/978-3-030-46736-4_2

for network, as the routers need to serve as content cache and enable routing based on nearest copy. ICN overcomes the drawbacks of current Internet by retrieving content based on its name rather than its location-dependent identifier, and adapt routing procedure that is centered on these unique names instead of IP addresses. It is a matter of concern whether this additional cost and complexity are justified or not. Moreover, it assures the same benefits if the ICN is deployed in an incrementally deployable manner. The research work mentioned in [3] represents the proof-of-concept model for an incrementally deployable ICN paradigm. It also explores the possible ways by which the ICN benefits can be achieved without any modifications to the network paradigm.

In the year 2000, D. Cheriton et al. [4] reported Translating Relaying Internet Architecture Integrating Active Directories (TRIAD) architecture of internetworking by explicitly adding a content layer over the IPv4 of TCP/IP protocol stack. It has aimed at enhancing characteristics of content routing, caching, and content transformation. TRIAD has brought a paradigm shift in Internet data delivery process. TRIAD is considered as the first formal proposal for content routing–based network or so-called information-centric network (ICN). After the proposal of Cheriton et al., many researchers have presented multiple architectures in support of ICN. An improved version of TRIAD with security provisioning is presented in the Data-Oriented Network Architecture (DONA). This work has a clean-slate redesign of Internet naming convention and name resolution to support searching of content based on the content name rather than host address. Similarly, structure of Content-Centric Networking (CCN) is proposed by Van Jacobson et al. [5]. The two most critical features of CCN are strategy and security which are included as new layers in CCN protocol stack. The strategy layer provides easy heterogeneous connectivity and the secure layer enhances security to contents rather than securing connections over transmission media as done in IP network. The Publish Subscribe Internet Technology (PURSUIT) [4] is proposed by Nikolaos Fotiou. Following this architecture, information is organized hierarchically, ranging from small data chunks to large documents or video files. Information is published using a unique rendezvous identifier (RId). Such named publications (i.e., content) are located either in the physical network or logical network (e.g., social network) and is called the scope of the published items. Likewise, each scope is identified by scope identifier (SId) and organized hierarchically. Contents are accessed using the RId and SId. Christian Dannewitz et al. [6] have proposed another ICN architecture, the Network of Information (NetInf). The NetInf architecture have emphasized on heterogeneous ICN deployment including traditional Internet access, core network configurations, data centers, as well as infrastructure-less networks. Recently, Named Data Networks (NDN) are proposed by Zhang Lixia [7] by enhancing the CCN architecture. In this work, authors have performed an analysis on CCN proposed in [8] and suggested some solution to issues involved in deployment and use of the CCN in operational networks and applications.

2.2 Representative ICN Architectures

There are many ICN architectures proposed by different researchers. In this chapter, few of those architectures are briefly described. A clear understanding of these architectures will help in understanding various protocols designed for ICN.

2.2.1 TRIAD Architecture

As reported in [4], the integration of content layer to ensure scalability of cached content and routing is the key concern of TRIAD model. The architecture supports path-based addressing mechanism extensively by using the shim protocol over IPv4 infrastructure. It also opens up the provision of communication using content name through network address translation (NAT). It integrates functionality for assisting mobility management, virtual private networks, policy-based routing, and source spoofing. It is compatible with IPv4, TCP, DNS, and other dominant Internet protocols of traditional network. The content layer enables a client to request data based on uniquely identified names, assigned to information in the network. The lower layers of the architecture can communicate content without much consideration about the content identification. The TRIAD content layer supports the underlying network by providing a network pointer for content retrieval. The fetching or modification may be accomplished through reference rather the transmitting the actual content. It is useful for large or medium-sized content where content transmission may be costly. The web Uniform Resource Locator (URL) is used as a format for content identification for the compatibility with the World Wide Web. Cookies are used as a network pointer to realize HTTP connections over TCP.

The content-specific level as stated in TRIAD is integrated in the Content Routers (CR) of the network model. The CR directs requests toward location of content. Locations may be the cache storing the content or the actual producer of information. Content caches transparently store information closer to the consumer to make faster delivery of content to the clients. The TRIAD architecture also includes a content transformer to transform content between different acceptable forms supported by underlying network characteristics and client specifications. In TRIAD, end users or the clients are identified by unique names similar to contents. From the content layer perspective, the producer or the cache of the searched content is one end point which is recognized from a URL. For interface perspective, the end point is a hierarchically arranged string used for a named content similar to DNS name. An address assigned to an interface may change over time, even during a session established by a transport layer. This name is used for both end point identification as well as authentication. Unlike IPv6 or IPv4, no global or persistence identifier exists in TRIAD for identification of end points.

The content layer of TRIAD is the heart of the architecture. It integrates naming and content routing. The routing procedure itself performs a name lookup on the content instead of implementing a separate functionality. The name lookup returns a forwarding track specification to the client who requests the content or information. The routing and directory mechanisms are strongly coupled in the router. The routing table is indexed by content names and mapping of names to the next-possible hop. All the routing information including end points and next-hop nodes are identified by name rather than address as in traditional network. Such name and name-mapping information are advertised by routing advertisements in the entire network. Any content producer may participate in the routing process by announcing the content name and size for which it is responsible. In each of the advertisements the source also specifies the distance to the content that it is advertising. A protocol namely the Name-Based Routing Protocol (NBRP) [9] in TRIAD performs routing by content naming. It works in the similar pattern with Border Gateway Protocol (BGP). However, as in BGP, NBRP uses names instead of IP, for each reachable information among autonomous systems. Routing information is distributed among CRs. The CR maintains table of content, name, and next hope (in name) to the actual location of the content. It makes the name directory and the routing table a logical single entity. It reduces overall computational complexity of CRs.

2.2.2   DONA Architecture

The architecture of DONA has targeted to redesign the naming system in TCP/IP network to fit into the content-centric environment. The model has changed the domain name services in IP networks by using flat and self-certifying naming convention to handle persistence and authenticity. It uses 128-bit flat name which makes informal identification harder and hence authentication easier. The name resolution handles availability and uses any cast name resolution process instead of the existing DNS. The model of DONA uses resolution handler (RH) to handle name resolution. The source of data is authorized with name resolution infrastructure and they need to publish their contents in the network to serve clients. To achieve high availability, the architecture uses route-by-name mechanism introduced by TRIAD rather than name lookup in IP network. An entity called principal is introduced to assign name to content.

A principal has a public–private key pair for security provisioning. Every piece or chunk of data is associated with a principal. The principal assigns a unique name to that data in the form P:L, where P is the cryptographic hash of the public key of the principal and L is the unique label of the data chunk assigned by the principal. DONA uses the same naming convention of P:L to identify services, hosts, and domains as well. In such cases, P represents the same cryptographic hash of the principal, but L represents the unique label of either service, host, or domain. Only hosts authorized by a principal with unique name can offer service or data. Likewise, this naming process ensures persistence because name does not refer to locations. A big challenge in naming mechanism of DONA is how to make such flat, long, and user-unfriendly names of data or services available to the end users. In conventional

TCP/IP network, DNS eliminates the need of remembering locations. However, DONA suggests that names should be made available to the end users via some external trustworthy mechanisms like search engines, private communication, recommender services, and so on. Users may keep these flat names in their own private namespace of human-readable names, which map onto these global and flat names for future reference. Moreover, DONA does not provide any reverse lookup mechanism [10].

Figure 2.1 depicts various data structures maintained by nodes in the architecture. These data structures are filled during registration process of a named content in DONA-based network. The diagram also shows the way a client requests data and how the architecture resolves this search into the actual location of the data. The architecture uses route-by-name mechanism in coordination with the resolution handler (RH) to convert flat names into actual locations. Two messages REGISTER(P:L) and FIND(P:L) are used to accomplish the task of name resolution. The "register" message is used by the handler to discover the route to the nearest copy. The "find" message is used by the client to request a data chunk or service. Every network domain or autonomous system is served by a single RH or hierarchically organized set of RHs. The RH information is known to all hosts within that region. Any host that wants to provide some data or service must register with the handler through respective principal. The registered name of the object (in the form P:L) is received from a principal. The handler then maintains a registration table with data name, next handler, and distance to the nearest copy. A "find" message comes from a client in the form P:L. On receipt of a "find" message, the handler looks up its registration table for the entry. If it finds one, then the "find" message is

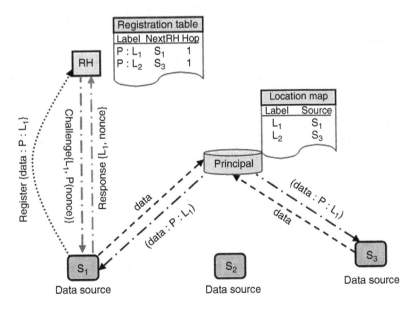

Fig. 2.1 Naming of data and registration in resolution handler in DONA

sent to the respective next RH as mentioned in the registration table. Otherwise, the message is forwarded to the next-level handler in the region or domain. Finally, the find message reaches either the nearest cached copy or the producer of the data. Once the "find" packet reaches the node with the data or the actual producer, it sends the data to the consumer. In the original version of DONA, it is stated that the data packets are communicated in the reverse direction of FIND packet, only when caching is implemented in all RHs. Otherwise, data may follow any optimal path from the source to the destination.

The DONA architecture provides the mechanism for authentication of both registration of data with the RH, and received data packets. To authenticate the registration of a data chunk, the RH sends a challenge by encrypting a nonce with the public key P as stated in the FIND (P:L). The client uses private key of principal who named the data to decrypt the encrypted nonce in the challenge. The decrypted nonce is sent back to RH as response. If response is correct, then the data is assumed authentic, and RH enters the mapping in the registration table. Data in DONA is self-certifying. A data chunk comes with a triplet (data, P, signature). Where P is the public key of the principal that provides name to the data chunk, and the signature is done by its private key at the time of naming, and could be verified by the public key P. On receipt of the data chunk, the receiver uses the public key included in the data to verify the signature. The signature verification ensures authenticity of the data. Hence, due to the property of "verifying authenticity" without any third party, the naming convention of DONA is called self-certifying.

The processing of FIND and data delivery process are shown in Fig. 2.2. The FIND packet is included between the IP and TCP header as a shim layer header. The FIND header carries the identifier of the content in the form of the name in a 40 bytes field as shown in the Fig. 2.3. The packet propagates to the top-level RH (Tire 1) if the named content is not available in the passing RHs on the path so far. If the information is not even available in the top RH, then the concerned handler generates an error message and sends it back to the generator of the find packet. If a RH finds the named content in response to a FIND packet, data is communicated to consumer as a standard transport-level response. However, transport protocols should bind services to names, not to addresses. Similarly, application protocols need to be modified to use names, not addresses. With implementation of DONA, many applications could be made simple. For application layer, URL information could be removed as the data is named in the lower layer, rather than in the application layer.

2.2.3 *Content-Centric* Network (CCN) Architecture

The CCN [8] treats content as primitive and retrieves contents by a name rather than its location as in host-centric IP-based network. It decouples location of data or content from its identity, security, and access. The original work of CCN has implemented the architecture's basic features with incorporation of resilience and

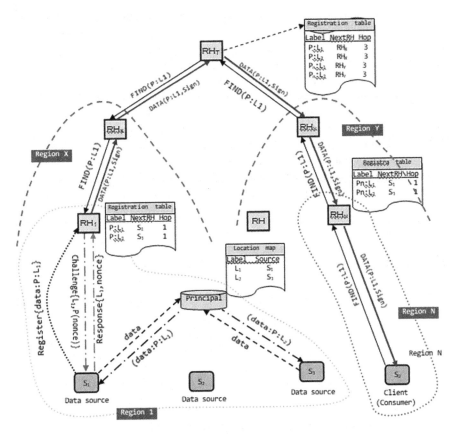

Fig. 2.2 Processing of FIND(P:L1) and DATA(P:L1,Sign) delivery

Fig. 2.3 Shim header communicating FIND packet used by DONA

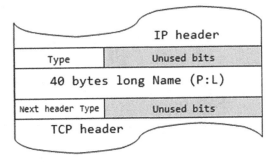

security. However, it preserves the design decisions to make TCP/IP simple, robust, and scalable. The communication is always initiated by the consumers. It uses two types of packets: interest and data. A consumer in need of data generates an interest packet by specifying the name of the content. It then broadcasts the interest packet over all its available links. However, the naming convention of CCN uses hierarchical name prefixes as */ThisRoom/laptop*, etc. Any node receiving the interest, replies to the interest by sending the data to the consumer if it has the requested data. If the node does not have the data, then it forwards the interest packet in all links. Because of identifying content with name rather than location, multiple nodes interested in the same content can share the same copy.

Figure 2.4 shows the processing of interest and data packets in a CCN architecture. Every forwarding node in CCN maintains three data structures: the Forwarding Information Base (FIB), Content Store (CS) and Pending Interest Table (PIT). The FIB is used to forward interest packet toward potential source(s) of matching data. It allows a list of outgoing interface(s) to send requested data. The CS keeps data packets for future use. Any interest packet with the content in CS is served from that node rather than forwarding to the actual source. If the cache store is full at the time

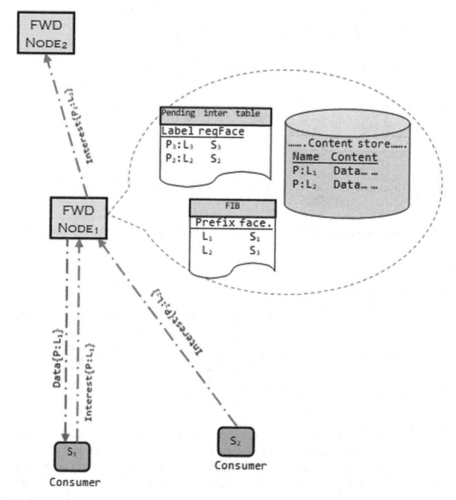

Fig. 2.4 Propagation of interest and data packets

of arriving a new content, then it uses either Least Recently Used (LRU) or Least Frequently Used (LFU) replacement algorithm to find room for the newly arrived data. It is different from IP buffering as IP uses Most Recently Used (MRU) replacement policy. CCN packets are idempotent, self-identifying, and self-authenticating. So, each packet is potentially useful to many consumers and hence maximizes the probability of sharing data packets as long as possible. The CCN recommends that the data packet follows the reverse path of the interest packets. It enables the forwarding nodes to cache content in the content store. For that purpose, it uses a data structure called PIT. The PIT keeps track of forwarded interests packets whose data is yet to arrive the forwarding node. Once the packet is arrived the PIT entries are erased. Entries which never find a matching data are eventually timed out and erased [11]. Figures 2.5 and 2.6 show the format of interest and data packet.

2.2.4   PURSUIT Architecture

In Publish–Subscribe Architecture, a source publishes information and a client subscribes to such content as their need. If the requested subscription is available, then the content is delivered to the requestor or the client. The Publish–Subscribe Internet Routing Paradigm (PSIRP) [12] was an earlier ICN model that proposed this architecture. The PSIRP is modified to design the publish–subscribe [P2] model by its successor project. Both these projects are part of the European FP 7 project. In this section, a brief description of the Publish–Subscribe model is presented as described in PURSUIT project.

The data unit in PURSUIT is called an information item. Each information item has a pair of identifiers called Scope ID (SId) and Rendezvous ID (RId). The Scope ID or SId specifies the group an information item and Rendezvous ID or RId represents the identification of each information piece within a group. Every single information available in the network must belong to at least one group or scope. The scope helps in policy enforcement and granting access rights for each group of information. Figure 2.7 shows a hierarchy of scopes as adopted in the architecture [12]. Naming of content or information follows flat scheme, identified by SId(s) and

Fig. 2.5 Format of interest packet

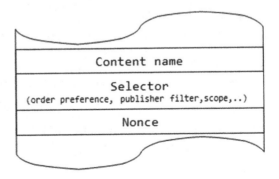

Fig. 2.6 Format of data packet

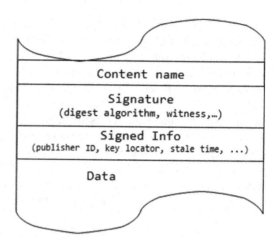

Content name
Signature (digest algorithm, witness,…)
Signed Info (publisher ID, key locator, stale time, ...)
Data

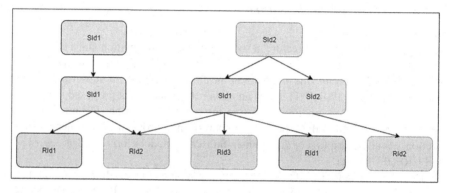

Fig. 2.7 PURSUIT information model

a RId. A name includes the path from the root of the tree to the leaf node, represented by RId. Same piece of information may have more than one name. For example, SId1/SId1/Rid2 or SId2/SId1/Rid2 both indicate the same piece of information Rid2 (Fig. 2.6).

There are three functional entities in the PURSUIT architecture. They are the Rendezvous Function, Topology Function, and Forwarding Function. Rendezvous Function provides a name resolution function to map a subscriber into a publisher. It also initiates the delivery of a content (information) to the requestor client. The Rendezvous Function is assigned to the Rendezvous Point (RP) on the Rendezvous Node (RN). RNs are structured in hierarchy through Distributed Hash Table (DHT) and responsible for managing different scope. The Topology Function deploys a routing protocol to collect the topology of its domain and exchange routing information with other domains for global routing. The Topology Manager (TM) in each domain controls the routing functions. The Forwarding Function is implemented on the Forwarding Node (FN). It forwards the information to a client using a label-based mechanism. It also uses the Bloom Filter technique to speed up the information delivery. The FN also caches information for later use by clients.

Figure 2.8 depicts the sequence of named data publish and request according to PURSUIT architecture [5, 13]. A source publishes information item (SId, RId) through the RN. The RN owns the scope of the named data. The scope owner RN may be a source local RN or a RN in different domain. The client specifies RId and SId in order to request for an information item. The client sends a subscription message through its local RN towards the scope owner RN using RId. The scope owner RN then sends a request to TM to generate a return path. The request is sent to the source of the information. It then uses source routing to deliver Named Data Object (NDO) to the client. The information item is forwarded via FNs, which uses Bloom Filter's Forwarding Identifier (FId) to decide where to send the packet. If requested information is previously sent to some other client, then the FN can directly forward the request to the source using a Bloom Filter's reverse path.

2.2.5   Network Information (NetInf) Architecture

The NetInf or the Network of Information architecture [6] treats data as the most vital entity and called it named data objects (NDOs). The network is assumed as a means of forwarding NDO for delivering it to the receiver or destination. The nodes in the network perform the operation of NDO forwarding activity. The NDOs are

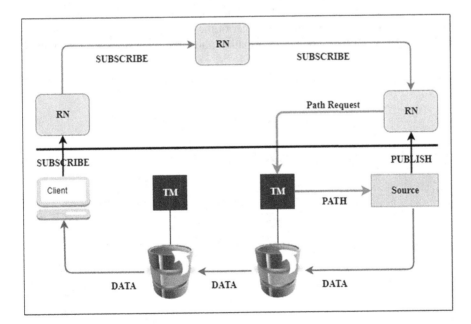

Fig. 2.8 PURSUIT, request and data flows architecture

location independent, and this architecture is accessed with its name rather than location, unlike host-based network. It enables ubiquitous replication and caching of NDOs in the network. The NDO has its name in common format with a common data structure. Forwarding process of NDOs is done using name-based routing mechanism. NetInf employs a flat naming convention for NDOs rather than topology-based or organizational approach. However, it has the limitation of lack of aggregating name-based routing or name resolution.

The diagram in Fig. 2.9 shows various components of NetInf with the propagation of various managerial packets. The architecture comprises of three different entities, the name resolution service (NRS) nodes, NetInf Routers, and the user nodes. It also uses a GET packet for locating data. The content is delivered to the clients through DATA packets. The NRS nodes facilitate name resolution to user nodes. The NetInf routers provide next hop information for name-based routing. The architecture uses a GET message to find a copy of the NDO in support of NRS. If the user node is aware of an NRS that holds the copy of the data, then it sends the GET request directly to the NRS. If it does not know the address of the NRS, then it sends name-based routing request to nearest NetInf router for finding the next hop router. Both the NRS and the adjacent NetInf routers are typically pre-configured at the user node. NetInf routers are placed inside provider networks to forward requests (GET message) and data internally as well as between domains to perform interdomain routing. The NRS-enabled node receiving a GET message may either send the NDO if it has a copy of it, or sends back a routing hint if it does not have one. The network hint is a name resolution service that helps in finding actual location of NDOs.

The GET request is forwarded hop-by-hop between nodes till a cached copy of the NDO is found or the original producer of the NDO is found. Any intermediate node, if it does not have enough information about the requested NDO, may then

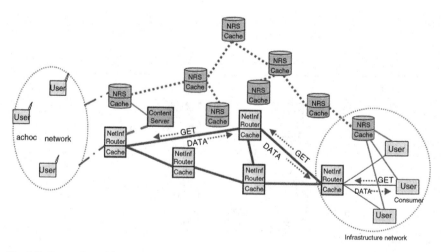

Fig. 2.9 Processing of GET and DATA and NRS tree in NetInf

perform a name resolution before forwarding the request which provides with a routing hint. Once the NDO is found, it is returned on the reverse path and same is cached for future use. Sometimes, a consumer of NDO may send a query to an NRS via GET messages to resolve the object name into a set of routing hints. Subsequently, the routing hints are used to retrieve the object via the underlying legacy IPv4 network. NetInf nodes allow interconnection of both ad hoc and infrastructure-based network through NetInf NRS and NetInf routers. Such routers are then connected to global Network of Information. The local NetInf router and higher-level NRS routers are updated with routing information publishing any locally registered information. The network providers run their own NRS to control the internal-network traffic and to reduce interdomain traffic. Each provider can set up a local hierarchy of NRS nodes that matches its network topology. Providers add NetInf-enabled in-network caches to their networks for performance enhancement. Off-path caches are connected to NRS nodes to retrieve information about local object popularity and to register cached objects.

2.2.6   The NDN Architecture

The Named Data Networking (NDN) architecture is reported in [8] to completely replace the IP architecture with generalized named objects. The proposal given by Van Jacobson and his team has emphasized on the functional components with comprehensive specification of routers functionalities. The thin waist architecture of TCP/IP model is redesigned with reference to the named object networking and components of the network layer is defined. NDN have demarcated the semantics of network services in terms of named objects so that data could be accessed by name rather than the source IP address as in host-centric network. The name in an NDN architecture could be used to identify all kinds of entities, like end points, data, and anything that contributes to communication system. This naming of objects also helps in content distribution and load balancing in a distributed environment. The architecture supports multipath forwarding and in-network storage of concerned named data objects.

The two types of packets, the Interest and the Data packets, are crafted by NDN architecture also similar to Content-Centric Networks (CCN), for communication of named objects. However, more detailed fields in these two packets are found in NDN description. For example, in Interest packet the guider field is included for scope and lifetime of the interest packet. In data packets, meta information of the content type and age of the packet is included. Both the interest and the data packets carry the name that identifies the piece of data that it carries. A consumer creates an interest packet with the name of the object (or data) that it is looking for. The routers in the network in turn uses the name to forward such interest packet toward the location of data. This location may be either source or producer of data or an intermediate router that caches a copy of the requested data. On receipt of interest packet by a node holding requested data, it returns requested data packet with the stated content. The data also carries a signature generated by the actual producer to ensure authenticity of transmitted data.

NDN provides a detailed description of the components of a router to accomplish the task of Interest and Data packet forwarding. It suggests to have three data structures: Pending Interest Table (PIT), Forwarding Information Base (FIB), and Content Store (CS). A packet forwarding module is also defined by NDN to decide the forwarding strategies. Description of each of these data structures are as follows. The function of a router starts with the reception of an Interest packet by a router. The PIT keeps track of all Interest packets that a router has forwarded but not responded with a Data packet yet. On receipt of an Interest packet, a router first checks its CS for the availability of the requested data. If exists, the router returns the Data packet to the sender of the Interest packet. If the requested data is not available, then the router looks up the data name in its PIT for a matching entry. If matching entry is found then it records the incoming interface of this Interest in the existing PIT entry, else it creates a new entry in the PIT. In the absence of a matching PIT entry, the router forwards the Interest toward the producer of the data. However, in case there is already an entry exists in the PIT, the Interest is not forwarded further. The PIT entry contains the data name requested by the consumer and the incoming interface(s) from which the Interest packet is coming along with the outgoing interface(s) through which the interest packet is forwarded. The function of the FIB is to record the probable producers of the data objects. While forwarding an Interest packet, the router consults the FIB to determine the outgoing interface(s). It acts as routing table in IP-based network. The router's adaptive Forwarding Strategy module assists the router to make use of the FIB in forwarding Interest packets. The FIB is filled by a name-prefix-based routing protocol. The CS defined by NDN is a temporary cache of Data packets, the router has received and keeps it for future references. Since in NDN, Data packets may be served from any of the locations where these are available, rather than only from the source as in IP network, the CS significantly reduces network overloading and reduces delivery delay by serving clients form intermediate places.

The flow in Fig. 2.10 shows the processing of Interest packet in NDN. The Interest packet traverses till either to the first router containing the requested data or to the producer of the data. The intermediate NDN router containing the data or the source returns the Data packets in response to the received Interest packet. The Data packet traverses in the reverse direction of the Interest packet flows. When a Data packet arrives at an NDN router, it first finds the matching PIT entry. Then it retrieves the list of all interfaces from which it had received the concerned Interest packet. Then the router forwards the data to all downstream interfaces listed in that PIT entry. The PIT entry for the Interest packet is then removed from the PIT. The NDN router also caches the received data in the CS for serving future consumers for the same data packet. Unless there is a loss of packets, one Interest packet results in one Data packet on each link. With this characteristic, NDN incorporates flow balancing in the network. Forwarding of Data packets are depicted in Fig. 2.11. All NDN routers forward Interest and Data packets based on the object names carried in such packets. The forwarding of Data packets toward the consumers is based on the PIT

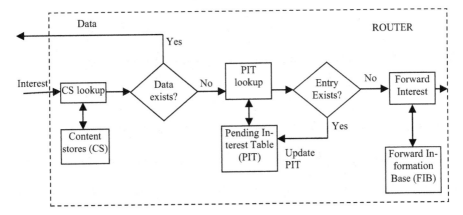

Fig. 2.10 Processing of Interest packet in a NDN Router

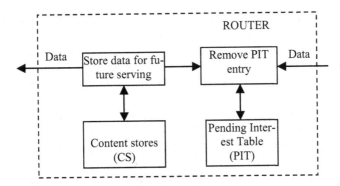

Fig. 2.11 Processing of Data packet in a NDN Router

entry at each hop set up at the time of receiving Interest packet. This Interest and Data packet exchange symmetry is termed as hop-by-hop control, and eliminates the need of including source or destination nodes' information in the process of data delivery.

The naming opaqueness in NDN allows different applications to choose its own naming convention. The architecture assumes a hierarchically structured naming scheme. For example, to identify a file called "file.ext" stored in the subdirectory under root may be named as /root/subdirectory/file.ext, where "/" is used to delineate different parts of the name. This hierarchical naming structure helps in representing the relations of data components with the location of the data. The names could be constructed deterministically to retrieve dynamically generated data. Deterministic approach helps consumers to construct the name for a particular data without having prior knowledge of the data name. To make global retrieval of data feasible, it should have a globally unique name; however, local communications may need only local routing to find matching data.

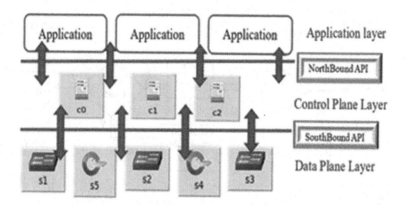

Fig. 2.11 Software-Defined Networking Architecture

The ICN, as discussed earlier in this chapter, is a wider representation of content (or information or data)-centric networking paradigm, whereas the NDN represents a specific architecture to realize the technology that defines various functional entities and operational behavior. The NDN (started around 2010) is a Future Internet Project funded by NFS and rooted from CCN (started around 2000) initiative led by Van Jacobson in collaboration with Xerox PARC. The prime goal of the NDN project is to design a running prototype of the prominent concept. There is a very narrow difference of NDN with CCN. The CCN refers to an earlier project of designing ICN architecture (initiated by Van started at PARC). It aims at designing the software codebase for the baseline implementation of this architecture to work as an overlay network with IP. Whereas the NDN emphasizes on a clean-slate design with complete change in its earlier architecture and removes the dependency on IP. Although it has the same origin with CCN, it has brought tremendous changes in the architecture including a clean slate deployment without depending on existing IP. That is the reason the CCN and NDN architectures have been taken up in separate sections in this chapter.

2.3 Comparative Analysis of ICN Architectures

In this section, a comparative analysis of the already discussed ICN paradigms is presented. Various parameters considered for comparison are the naming mechanisms used, how name resolution and content routing take place in one paradigm, which caching strategy it supports, whether it supports mobility of publisher/subscriber, and which security provisioning mechanisms have been incorporated in the respective paradigm. Each paradigm is explained with respect to these key parameters in Sect. 2.2. Table 2.1 represents the comparative analysis for the same followed by Table 2.2 which discusses the advantages and disadvantages associated with each ICN paradigm.

Table 2.1 Comparative analysis of ICN paradigms

Model	Naming	Name resolution and data routing	Caching	Mobility	Security
TRIAD 2000	DNS-based naming schemes (hierarchical)	A protocol named Name-Based Routing Protocol (NBRP) [9] in TRIAD executes routing through content naming	On-path caching. Content router caches transparently cache data nearer to the end user to make faster data delivery to clients	The key method to enable mobility is the adding and deleting of redirects inside mobile host's home directory and the mobile host registration inside guest realm	The security provisioning mechanism has not been incorporated
DONA 2006	Flat namespace, consisting of label and principal part	Hierarchical organization of Resolution handlers. Coupled: A source path is generated during resolution. Decoupled: Resolution handlers will return the network address	On-path caching is enabled at resolution handlers. Off-path caching needs extra registrations	Subscriber mobility is supported through new requests. Mobility of publisher needs extra registrations	Principal part in name is hash of public key. Label part in name can be hash of immutable data object
CCN 2007	Hierarchical naming	Name-based routing as per the routing paths mentioned inside FIB table	On-path caching is supported. Any other caching strategy can be integrated	It uses the protocol named *Listen First Broadcast Later* (LFBL)[14] to implement support for mobility in opportunistic/ ad hoc networks	The data publisher provides security features by cryptographically signing each content packet
PURSUIT 2011	Flat namespace, consisting of rendezvous and scope part. Scopes may be arranged in hierarchical fashion	DHT-oriented rendezvous network matches Content subscriptions to its publications. Scopes will be used to restrict the resolution /publication of content. Decoupled: Forwarding and topology management functions are isolated from rendezvous	On-path caching becomes difficult because of decoupled mode of operation. Off-path caching needs extra registrations	Subscriber mobility is supported through new requests. Mobility of Publisher needs Topology manager updation.	Packet level authentication is there for each packets. identifiers can be self-certifying optionally

(continued)

Table 2.1 (continued)

Model	Naming	Name resolution and data routing	Caching	Mobility	Security
NetInf 2013	Flat namespace	Name resolution and content routing can be either coupled or decoupled module or hybrid in this paradigm. Coupled: A routing protocol is utilized to advertise data object names and fill the *Content Routers'* routing tables Decoupled: A *Name Resolution System (NRS)* is utilized to do mapping among object identifiers to locators that can be used to find the related data objects, like IP addresses Hybrid: In this mode, the NRS gives *routing hints*, which is, partial locators that can guide a GET message in one or more routes where much data related to the needed content object may be found	Apart from on-path caching at the CRs, it also visualizes the deployment of large-scale data object caching and replication methods with cooperation to NRS	Mobility of host is enabled by having the NRS that keeps track of topological data for each registered host node	This paradigm visualizes a full-fledged Security model that contains security of name, data integrity, authentication, authorization, confidentiality, and provenance
NDN 2014	Hierarchical namespace, it can contain publisher-specific prefix	Routing protocol is used for dissemination of name prefix data. Coupled: Routing state for data is created at content routers during request of packet propagation	Supports on-path caching at content routers. Off-path caching needs extra routing information	Mobility of subscriber is enabled through new requests. Publisher mobility is supported through Interest flooding protocol	Signatures are included in each packet. Certification chain follows identifier hierarchy.

Table 2.2 Comparative analysis of ICN paradigms

Model	Pros	Cons
TRIAD 2000	It resolves the problem of scaling data distribution by defining a content layer that directly enables efficient data routing, transparent caching, and data transformation	The proposed paradigm has not emphasized on security provisioning schemes
DONA 2006	Identifiers in DONA are self-certifying, that is, they permit the subscriber to verify that the received content matches with the name required	The DONA paradigm does not provide any mechanism for reverse lookup [10]
CCN 2007	• CCN packets are self-identifying, idempotent, as well as self-authenticating • Hierarchical namespace is used to attain good routing scalability	The overall maintenance of forwarding strategy may get increased due to more number of data structures maintained by content router
PURSUIT 2011	• IT also uses the concept of Bloom Filter technique in order to speed up the data delivery • PURSUIT provides the *Packet-Level Authentication* (PLA) method for encrypting as well as signing each packet	Name resolution process can be time consuming, specifically since the DHT routing does not follow the shortest routes among the communicating nodes
NetInf 2013	• The network providers will run its own NRS in order to control the internal network congestion and to decrease the interdomain congestion • This paradigm can be deployed as an additional layer on top of the existing network paradigm, which simplifies the migration of applications to the new paradigm	It lacks with an efficient content-retrieval mechanism which produces less content retrieval latency
NDN 2014	NDN paradigm has been introduced with potential for a wide range of advantages like content caching to decrease congestion and increase delivery speed, simplified configuration for network devices, and building security methods inside network at the data level	To make global content retrieval feasible, it must possess a globally unique identifier

2.4   Software-Defined Networking in ICN

Software-Defined Networking (SDN) is an emerging network architecture with the separation of network control functionality from forwarding module. The migration of formerly tightly coupled control with forwarding strategy in individual network devices is converted into programmable distinct components. This accessibility of computing devices enables the infrastructure to appear as an abstract unit to applications and network services. It opens up the possibility to treat the network as a logical or virtual entity. In SDN, the network intelligence is (logically) centralized in software-based SDN controllers. These controllers maintain a universal view of the network. It enables the network to appear as single logical switch to the application interface and policymakers. SDN supports vendor-independent control over the network to enterprises and carriers. It greatly simplifies the network design and operation process form designer's perspective. SDN also simplifies the functionalities of network elements by omitting the need of understanding and processing large number of protocol standards as it merely accepts instructions from the SDN controllers without knowing much detail about the scheme.

Traditional network architecture cannot meet the current demand of enterprises, carriers, and end users. The SDN have come as a means of transforming networking architecture by allowing programmers to tune the required parameters for data transmission. Consequently, enterprises and carriers gain unprecedented programmability and network control to build highly scalable, flexible networks. The OpenFlow-based SDN is currently being rolled out in a variety of networking devices and software. It possesses substantial benefits to both enterprises and carriers. Few of such benefits are [14] as follows:

- Easy management and control of networking devices
- Improved automation and management through common APIs
- Rapid implementation of new innovation through the ability to configure individual devices
- Increased network reliability and security via centralized and automated network devices
- More granular network control at the session, user, device, and application levels
- Better end-user experience due to centralized network state information to seamlessly adapt network behavior to user needs

SDN is a dynamic and flexible network architecture with the ability to program devices as per the need of the end users and the network. It can evolve today's static network into an extensible service-oriented data delivery platform with immediate response rapidly to changing business, end user, and market demands.

2.4.1 Scope of Integrating SDN into ICN

ICN is the most promising network technology that may eliminate majority of the limitations of host-centric communications. As mentioned above, Named Data Networking (NDN) is one of the most credible architectures for effective deployment of ICN. However, till date, there is lack of effective cache replacement strategy and routing mechanism in NDN. SDN, on the other hand, is another dimension of recent notable research trend in networking. It transmutes the traditional networks into a service delivery architecture by decoupling control and data plane. This decoupling isolate network infrastructure form user applications. It, in turn, facilitates the easy adoption of new network architectures in a more flexible way. With the advent of SDN, researchers are trying to integrate it with different network implementations. It is found to be integrated into ICN also. The synergy of ICN's efficient data dissemination and SDN's flexible network management is promising to design a fully controllable framework for efficient data communication. Primarily, the recent proposals are coupling SDN with caching and routing in ICN to take the advantages of data and control isolation in improving scalability and dynamism.

In line with the SDN, ICN can also be visualized as the integration of two functions. One part of the functionalities includes the awareness of user demand (or popularity of data) and the other part forwards data chunks. The first part, which keeps track of data popularity based on metadata, may be considered intelligent. This intelligent part assists data forwarding part by providing the information of probable path of data dissemination. The data forwarding part does not have intelligence and simply forwards the data other as per the direction of intelligent part. This separation of functionalities in ICN is similar to the separation of control and data plane in SDN. Hence, it is well justified to integrate SDN into the framework of ICN [18, 19, 20, 21].

The integration of SDN into ICN is expected to bring the following benefits:

- ICN has several distributed components used for naming, caching, routing, and security provisioning. However, ICN does not have a separation mechanism of managing such distributed components. SDN may support ICN by providing an efficient management framework to control such distributed components of ICN.
- The ICN is an upcoming technology and continuous development is in progress. Hence, an SDN-enabled framework will certainly help researchers to design new protocols and architectures for ICN without much worry about its deployment.
- ICN emphasizes on data dissemination based on content names. Hence, popularity of data has utmost contribution to the success of ICN. By separating intelligent part of ICN and integrating with SDN control plan may certainly ignite the scalability by globally collecting information well in advance before data request arrival.

This chapter is aimed at designing an SDN-enabled framework for ICN architecture that will leverage the scalability of the later. It is primarily focusing in facilitating caching and routing through SDN in ICN environment. The proposed framework will help in the deployment of an ICN architecture within the framework of SDN with optimal modifications in the architectural and functional components.

2.4.2 The SDN Architecture in General

The recent trends in Software-Defined Networking (SDN) significantly encourage communication technologies, network platforms, and programming-based networking. Traditional networking is not as suitable as SDN, because that has lesser capability to handle heterogeneous demand of networking. Such capabilities make SDN prominent networking technology. SDN have several advantages over conventional networks as mentioned below:

1. Decoupling of data plane and control plane with programmability

2. Conceptually centralized controller configuration to accelerate dynamic network with heterogeneity
3. Easy software upgrade with reference to cross layer functionality
4. Sufficient environment testing with quick deployment, etc.

In SDN, network intelligence is handled by a logically centralized controller that is also known as network stack or network brain or simply SDN controller. However, another important component in SDN is switch which is responsible to forward data from one host to another.

SDN architecture can be viewed in terms of three layers (see Fig. 2.11). In this architecture lower layer is entitled as data plane, middle layer represented as control plane, and the higher layer is known as application layer. The data layer is responsible to forward data flow, while control plane is responsible for decision-making functionality. SDN is applicable in data centers and cellular networks. Applicability of SDN is not limited to above two areas only. Therefore, the last layer entitled as application layer of SDN consists of lots of applications in different areas.

As in Fig. 2.11, the applications can access a network through northbound Application Programming Interface (API). Northbound APIs are high-level programming interface to translate application requirements into lower-level service requests. Whereas control plane and data plane connectivity APIs are southbound. There are various southbound APIs in which one of them is OpenFlow protocol. It is widely acceptable and southbound API providing specifications for OpenFlow-enabled forwarding devices and communication channels is deployed. In that reference, OpenFlow provides three pieces of information to the respective controller; first is event-based messages; second is flow statistics, and third is packet-in messages. This southbound API is an important part because it affects performance and scalability of SDN. The popular southbound API OpenFlow leads in terms of CPU loads. Another variant is Cbench which works fine with system configuration and environment. Performance evaluation in SDN is based on throughput, latency, bandwidth, and workload.

2.5 Summary

This chapter describes all the architectures proposed for implementing the ICN from its interception. The functional and architectural advancement in all these models are stated to give a clear understanding of the technological operation. The key features of such architectures are also discussed. These architectures are different forms of various aspects like naming, caching, security, etc. However, their main objective is to provide access to data based on name. Many of the protocols are supporting the integration of content routing and name resolution along with mobility. All ICN architectures are designed to cope up with the needs of current Internet users. The following table summarizes each ICN architecture with its name,

Table 2.3 Facts about various ICN network architecture

Architecture, year	Associated projects	Major contributions	Packets and data structures used
TRIAD, 2000		• Adding content layer on the top of the IP layer in TCP/IP protocol stack • Named contents for unique identification	
DONA, 2006	• DONA project at UC Berkeley	• Defined a formal architecture of TRIAD • Enhances security provisioning to TRIAD • Defines a resolution handler to handle data search with name	• Resolution handler • FIND • REGISTER • DATA
CCN, 2007		• Added strategy layer to connect heterogeneous networks • Security provisioning to contents rather than transmission path • Defined a three data structures for forwarding nodes	• INTEREST • DATA • FIB • PIT • CS
PERSUIT, 2011	• The PURSUIT project • Publish Subscribe Internet Routing Paradigm (PSIRP) project • Both are funded by the EU Framework 7 Program (FP7)	• Introduced the notion of rendezvous identifier to identify content • Introduced notion of scope identifier to identify location of content • Proposed hierarchical arrangement of content and scope for efficient access	
NetInf, 2013	• European FP7 4WARD project • Scalable and Adaptive Internet Solutions (SAIL) project	• Supports heterogeneous network architecture including infrastructure less network	• GET DATA NRS node NetInf Router • Defined ad hoc support
NDN (CCN Revisited), 2014	NDN Project US NSF (Future Internet Architecture)	• Suggested deployment and use of CCN in network and applications • It does not propose any new architecture, CCN architecture is explored in this work	

proposed year, major contribution of each, related funded projects, as well as data structures and packets used inside it. This table will help as a quick reference for the survey of ICN architectures (Table 2.3).

References

1. Pentikousis, K., et al.: Information-Centric Networking: Baseline Scenarios. No. RFC 7476 (2015)
2. Teemu, K., Mohit, C., Chun, B.-G., Ermolinskiy, A., Kim, K.H., Scot, S., Ion, S.: A data-oriented (and beyond) network architecture. SIGCOMM Comput. Commun. (Rev.). **37**(4), 181–192 (2007)
3. Fayazbakhsh, S.K., Lin, Y., Tootoonchian, A., Ghodsi, A., Koponen, T., Maggs, B., Ng, K., Sekar, V., Shenker, S.: Less pain, most of the gain: Incrementally deployable ICN. In: ACM SIGCOMM Computer Comm. Review, pp. 147–158 (2013)
4. Cheriton, D., Gritter, M.: TRIAD: A New Next-Generation Internet Architecture (2000). Online http://www-dsg.stanford.edu/triad/
5. Fotiou, N., Polyzos, G.C., Trossen, D.: Illustrating a publish subscribe internet architecture. J. Telecommun. Syst. **51**(4), 233–245 (2011)
6. Dannewitz, C., Kutscher, D., Ohlman, B., Farrell, S., Ahlgren, B., Karl, H.: Network of information (netinf)–an information-centric networking architecture. *Comput. Commun.* **36**(7), 721–735 (2013)
7. Lixia, Z., Alexander, A., Burke, J., Jacobson, V., Claffy, K., Crowley, P., Papadopoulos, C., Wang, L., Zhang, B.: Named data networking. SIGCOMM Comput. Commun. (Rev.). **44**(3), 66–73 (2014)
8. Jacobson, V., Smetters, D.K., Thornton, J.D., Plass, M.F., Briggs, N.H., Braynard, R.L.: Networking named content. In: 5th International Conference on Emerging Networking Experiments and Technologies, pp. 1–12. CoNEXT, New York (2009)
9. Gritter, M., Cheriton, D.: Name-Based Routing Protocol Specification, in progress (1999)
10. Koponen, T., Chawla, M., Chun, B.-G., Ermolinskiy, A., Kim, K.H., Shenker, S., Stoica: A data-oriented (and beyond) network architecture. SIGCOMM Comput. Commun. Rev. **37**(4), 181–192 (2007)
11. Carofiglio, G., Gallo, M., Muscariello, L.: Joint hop-by-hop and receiverdriven interest control protocol for content-centric networks. ACM SIGCOMM Comput. Commun. Rev. **42**(4), 491–496 (2012)
12. PURSUIT Project Includes Publications and Technical Reports of PURSUIT Architecture and Design. Online available at http://www.fp7-pursuit.eu/PursuitWeb/
13. Xylomenos, G., Ververidis, C., Siris, V., Fotiou, N., Tsilopoulo, C., Vasilakos, X., Katsaros, K., Polyzos, G.: A Survey of Information-Centric Networking Research. IEEE Commun. Surv. Tutorials. **PP**(99), 1–26. http://ieeexplore.ieee.org/document/6563278/
14. Meisel, M., Pappas, V., Zhang, L.: Ad hoc networking via named data. In: ACM MobiArch, Chicago, pp. 3–8 (2010)

Chapter 3
Naming for Unique Content Identification

3.1 Introduction

An information-centric network emphasizes on what data is being transferred instead of which elements of networks are transferring data. In the context of ICN, uniquely named objects are searched in place of uniquely numbered hosts as in the traditional Internet. The diversion from host-driven to data-driven concept help is effectively making use of distributed data inside a network. Recently this transition in paradigm has attracted many researchers to investigate and explore the same. Even different research projects have been initiated across the globe to advance this field [1]. The primary research concerns are naming, caching, and routing, which are the major open challenges in information-centric networks domain. The purpose of this chapter is to explore the concept of naming in ICN, its significance, distinct naming approaches proposed till date with associated major ICN research projects, comparative analysis of the existing naming mechanisms in ICN, and new proposed research direction toward efficient naming scheme in information-centric networks.

The current paradigm of the Internet has been developed in the year 1960 to connect several computing nodes geographically dispersed. The TCP/IP emerged as a solution-cum-communication paradigm in which hosts can set communication paths among themselves to communicate with one another. It was best suited for several client–server utilities like FTP, HTTP, SMTP, and telnet. Still host-oriented paradigm of the Internet is not capable enough for recent bandwidth-sensitive applications of end users. This fact needs a transition in the paradigm of the current Internet from host oriented to data oriented. Since 2007, HTTP videos are ruling the Internet traffic because of the arrival of various video streaming platforms for user-produced data (Google and YouTube videos) and video-on-demand portals (IPTV and Netflix). As per the 2012 report of the Cisco visual networking index, around 26 exabytes of total Internet traffic is getting produced every month, out of which more than 55% of traffic is generated due to videos [1].

© Springer Nature Switzerland AG 2021
N. Dutta et al., *Information Centric Networks (ICN)*, Practical Networking,
https://doi.org/10.1007/978-3-030-46736-4_3

The Internet has emerged in an ad hoc fashion. Different modules were inserted into Internet to meet new needs of users as they came. Domain Name System (DNS) is one such example that provides the mechanism for the resolution of names by mapping URLs into IP address. However, it does not provide support for movement and replication of content as well as awareness about location [2]. To support location information, content distribution network (CDN) has come into the picture. In the same way, peer-to-peer file-sharing applications, like BitTorrent, were developed for the multiple source data retrieval, content replication, and rapid distribution of data. All these schemes have given their contribution to the improvement of data access on Internet. Still, in the general case, they work as an overlay and do not utilize the information of the base network topology to attain optimal efficiency.

These drawbacks have given the motivation to researchers to work for another paradigm of future Internet [3]. ICN is one of such paradigm, best suited for the future of Internet. So the major goal of ICN is to shift the existing host-centric communication paradigm to data-oriented paradigm. It depends on naming which is location-independent, name-driven routing of messages, and in-network caching in order to effectively distribute the data inside network. Though ICN attracts a wide range of interests from the research community, the amount of work done in this field is still at the beginning stage. Various research challenges need to be resolved to make ICN the future of Internet's paradigm, like name-driven routing, persistent and secure naming, in-network caching, resolution of name, privacy and security of data, on-demand duplication, distribution of data, incremental implementation capacity, and backward compatibility. As mentioned earlier, naming, caching, and routing are three major areas of research in ICN paradigm. Numerous ICN projects have proposed different mechanisms for naming, caching, and routing. The target of this chapter is to carry out an in-depth discussion about various naming schemes in each ICN project. Furthermore, recent naming mechanisms for ICN paradigm with its strong and weak aspects are also covered. A comparative analysis among all of these methods and proposed new insight toward an efficient naming scheme in ICN is also presented.

3.2   Naming System Design

To uniquely locate the resources in ICN, an efficient and unique naming mechanism is a must. This includes the naming of all the network elements including the content itself. Based on the content name, routing is supposed to perform. Hence, the performance of routing also depends on how well the naming mechanism has been built and how much time it takes to resolve given names. Each name should be a unique identifier to any given network resources along with location independency in nature. The purpose of information-centric network is to focus on "what" instead of "where." The existing Internet allocates IP addresses to each network node and interfaces [4]. But ICN gives unique names to data. In reality, allocating unique, efficient name to each content chunk in-network is a big challenge. Moreover, the naming style must be self-sufficient against any possible network attacks. It should also be lightweight in terms of required processing power so that it can get adopted by ICN-based IoT infrastructure as well. There are several existing naming mechanisms adopted by several major ICN projects to assign a name to data.

The adopted naming convention in ICN is inspired by the DNS functionality of the traditional Internet. However, the methods used in DNS cannot be adopted to the ICN directly due to its inherent limitations. The existing DNS manages the administration overhead that occurred due to the increasing count of total host nodes. If several nodes on the Internet are too large, there is a need for a resolution system that is easy to manage in terms of administration, scalability, and distribution. This resolution scheme maps the domain name or hostname to a corresponding IP address. This makes the process of data retrieval from remote nodes or access to remote nodes easier as users just need to enter human-readable names. Though successful adaptation of DNS on Internet since a long duration, the existing domain name system suffers from various technical restrictions to face upcoming security needs. For example, security against different possible network attacks such as denial-of-service attacks (DoS attacks) and DNS cache poisoning [4], etc. Apart from this, DNS system has no information regarding client location. So, DNS cannot answer the query related to location of the nearest present sources. The major reason for this problem is with the base of the paradigm. The lack of an efficient content naming scheme that is independent of forwarding, routing, as well as storage methods introduces many issues for ICN. A naming method may be flexible enough to retrieve data from various sources with security information integrated into the name of content. There are various naming mechanisms available in the literature that emphasize distinct aspects to achieve higher network performance.

As the ICN naming and identification is influenced by DNS, hence a better understanding of DNS helps in designing a naming scheme for ICN. Here a brief description of DNS is described for a better understanding of readers. To ease the resource access on the Internet by its associated names, the name resolution process is depicted in Fig. 3.1. The figure specifies that whenever user requests a URL, it includes the domain name as well as the resource name. The request is sent to the desired content server. But it is first sent to DNS server for finding the corresponding IP address of resource container server. The transformed request is sent to a web server that holds the desired data [5]. As the Internet grows in size and the bandwidth consumption shifts from email and remote access to bandwidth-sensitive applications like YouTube, limitations are found in DNS in such scenarios. Moreover, "lack of location information" of the existing DNS system makes it unreliable for ICN. DNS cannot report the nearer data sources available inside network. The alternative approach CDN can make such information available up to certain extent. It works by performing redirection of DNS queries to their nearest data servers from where needed data could be fetched. Still, CDNs are limited to a group of users that pay to this service that cannot be served by DNS natively [5].

To solve the limitations of the current Internet paradigm, numerous ICN schemes have been introduced for the next generation Internet. It includes the proposals CCN, DONA, and NetInf. All these emphasize on efficient data distribution by

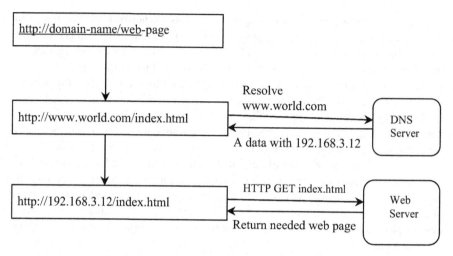

Fig. 3.1 DNS name resolution

unique content naming. The majority of these schemes have suggested indirectly or indirectly emphasize locator/identifier splitting. Data is recognized by its related ID value which is location independent. This ID is not useful for routing in ICN with existing methods like IP. So, how to fetch information depending on these ID values is a problem that is required to be solved by ICN paradigms. In the next section significance of the same is explored along with some existing naming mechanisms in ICN.

3.3   Importance of Naming

The information-centric networks overcome the issues present in the existing Internet paradigm including the current domain name system to map user-friendly domain names to location-dependent IP addresses. The problems and security vulnerabilities present with the current domain name system and its working principles are discussed in the earlier part in this chapter. The objective of the information-centric networks is to perform all the network primitives by keeping information on center. That means all the operations including caching, routing, naming in ICN revolves around content. So basically, information-centric networks can be differentiated from existing Internet in three major aspects like naming data, interdomain data routing, and the location of the narrow waist with an information-centric network-oriented Internet [6]. As shown in the previous section that in information-centric networks, each end user sends a request for the desired content with the help of its name while not know location of the content. The significance of naming is to be able to uniquely identify the content present inside network. Based on this unique identifier, routing is performed instead of IP addresses. The in-network caching is performed with the help of this identifier only. So, to name each content chunk in such a way that caching and routing are performed efficiently is also a major challenging task. And security comes at a front role in this case. It is expected that any data-oriented security approach must satisfy the below-mentioned features [6].

- The end user should know the name of the data and its type. It means user should map among a description of what they want and its related ICN identifier, which is resolved by the system itself.
- The user must be aware of the public key of the data provider for any specific data so that the user can verify the integrity/correctness of the received data.
- The information-centric network should permit binding between the content name and the public key of its provider. Due to this many attackers are prevented from doing false registration of data. If the system does not perform this binding, it may introduce several possible network attacks like denial-of-service (DOS) attack.

So far, discussion carried out is about the fact that the naming of content lies at the core of ICN paradigm as it is the basic foundation to efficiently perform two major primitives of ICN like caching and routing. In the context of naming, information-centric network has two major naming approaches. The first approach is similar to the current domain name system namespace and makes use of hierarchical names that are human-readable too [7]. The feature of human readability meets the first need partially and the hierarchical concept supports scalability. Various schemes may permit the user to get the public key like webs of trust, public key infrastructures, and personal contacts. But in the case of information-centric networks, to get information about public keys needs a globally approved public key infrastructure to bind identifiers to public keys. In opposite to that, the other naming scheme considers self-certifying identifiers. In this scenario, the public key is bound to the identifier itself, thus the information-centric network system need not rely on any public key infrastructure. Still, these identifiers are not in human-readable format. Therefore, users need to take the help of other schemes like webs of trust, portals of search engines and personal contacts, etc. to find the identifier of the needed data.

To satisfy the interest packets for any data, the information centric network system needs to forward these requests or interest messages. There are various name-based routing mechanisms present in the literature of information-centric network. The various information-centric network projects differ majorly on the routing approach adopted in the realization of the technology. The content-centric networking (CCN) utilizes the existing interdomain routing as a basic routing module and builds the name-oriented routing on top of border gateway protocol (BGP). CCN is one of the paradigms of information-centric networks. Others like data-oriented network architecture (DONA) utilizes the BGP model but it follows its own name-driven routing method. And some other ICN paradigms, like Publish Subscribe Internet Routing Paradigm (PSIRP), build their paradigm for interdomain routing. So there is a major role/significance of naming in information-centric networks to perform name-based routing efficiently [6].

ICN considers data as a first-class citizen. Each content chunk has assigned a unique identifier. The routers in the ICN forward data depending on the data identifier instead of the data locations. This leads to advantages like network caching, support for mobility, and dynamic routing of requests. In this case, the selection of naming method is also a major design consideration in ICN which finds the functional isolation among application layer and network layer. There are two major categories of naming schemes, like hierarchical and flat naming, that are discussed in detail in the upcoming section with comparative analysis. In the case of hierarchical naming, the network layer is aware of the meaning of the hierarchical names. The management of the namespace is the responsibility of the network layer. ICN paradigms that use a flat naming scheme leave this responsibility to the application layer. The choice of naming method affects the scalability and performance of the ICN in the context of the size of namespace and routing table as well as routing efficiency.

End user wants to retrieve data present in the network irrespective of its provider, location, and the method to fetch the same. In IP-based scenario, the entire process of data retrieval contains two major steps: (1) resolution of name: this maps the URL entered by user to its IP address (current location) and (2) location-driven routing: to fetch the data present inside network [8]. The identifier resolution is referred as domain name system lookup and done in the application layer before the routing takes place. This binding at initial stage leads to wastage of network resources in case there exist multiple alternate data sources in the network. This results in high content retrieval delay because users cannot retrieve data even if the copy of it exists in nearby location. Apart from this, the above method cannot take the benefit of in-network caching because the network is unaware about the presence of multiple data sources that can satisfy the same content request.

To overcome the issues presented above, ICN considers data as a first-class citizen. In information-centric networks, many new network paradigms have been evolved like named data networking (NDN), network of information (NetInf), data-oriented network architecture (DONA), etc., and all these paradigms convey the same objective of name-driven, location-independent interaction but has distinct approaches. The important aspect of these paradigms is the use of a data identifier to retrieve needed data. As the network starts to understand the data, the above-mentioned solutions can incorporate name resolution service inside network layer either off the path or in the path. Now each router in ICN can perceive and choose the path to route the content requests (called late binding), the network identifies better data sources inside network. This decreases the delay and burden on the data providers as well as on the network itself. The ICN paradigms also support network caching that helps to optimize network efficiency due to its data cognizant ability [9].

To support naming space for data is not as easy as it seems in the case of the modern network routers with memory capacity in the order of gigabytes. The amount of data is predicted to be one to two orders of magnitude greater than the existing IP address space. The fact is that the namespace for content can be greater than the predicted one. The selection of the naming method is a major and important thing for each ICN paradigm because it directly affects the scalability of network, routing efficiency, network management, and several other parameters of the entire paradigm. The approach for naming also acts as a major differentiator between the information-centric network paradigms.

The naming mechanism in information-centric networks is categorized majorly into two types: (1) hierarchical naming and (2) flat naming. Both the naming schemes have their pros and cons related to them. The selection of anyone depends on the requirement for the performance. Hierarchical naming gives the advantage of aggregation that leads to a reduction in the size of the routing table and computation as well as storage needs of the router. It also supports network scalability and gives rich semantics. In the case of flat naming, the lookup procedure for names has higher efficiency in comparison with hierarchical naming.

The effectiveness of network paradigms can be measured with the help of hierarchical and flat naming using distinct performance metrics like aggregation ability, routing efficiency, and semantics. Though these parameters are not fully dependent, they support each other to achieve the desired goal. For example, the absence of aggregating ability can affect the routing efficiency because of the larger routing table size. It can be affected by duplication and mobility of data in the network, and this leads to the increased size of routing table even in the case of a hierarchical naming. The management of name space can be affected by the pattern of communication and semantics. And this results in distinct sizes of tables.

Therefore, it is desirable to have an in-depth investigation and detailed study about selection of an appropriate naming mechanism for information-centric networks as this turns into a key parameter in their performance. In the next section, there is an in-depth discussion of various naming methods proposed so far with examples and detailed analysis about their strengths and weaknesses in context of information-centric networks.

3.4   Key Design Issues and Choices for Information/Content Naming for ICNs

To name a data object is an integral and important part of information-centric network, same as naming a host on the current Internet. ICN needs unique identifiers for each named data objects since these identifiers are used for recognizing data objects irrespective of their physical location. Several issues need to be addressed for the efficient naming design implementation in information-centric networks. In this section, there is a discussion of the same in the context of information-centric networks.

- To name static content objects can be done by using data hashes as a part of data identifiers so that data publishers can compute the hash value over requestors and existing content objects. Any node of the ICN can perform validation on the identifier-data binding by recomputing the hash value and comparing it with the identifier component.

- There are use cases where the identifier has to be produced before the data object is generated. In this case, it uses dynamic approach for naming content objects. Let's take example of live streaming when a data publisher needs to make the data stream available by doing registration of data stream chunk identifiers inside network. An approach like hash-based identifiers along with indirect binding as explained above may be a good example.
- The privacy protections of the requestor can be an issue that needs to be addressed. This is the direct outcome of the named content objects approach: If the network can see the request packets and response packets, it can also maintain request packet history/logs specific to network segments or for a particular user. This is not a desirable approach as identifiers are expected to be lasting for a long time. That means though the identifier itself does not convey much information, it is assumed that the identifier can be used to fetch the related content objects in the future.
- The procedures for versioning and updating named data objects can be challenging due to its contradiction with fundamental assumptions of ICN: If a named data object can be copied and stored inside in-network caches for future retrieval purpose, identifiers have to be lasting over long duration and binding between name and data must not alter; changing the data without producing a new identifier is not possible. The possible solution can be versioning but it needs a naming mechanism that assists it. It also needs a way for the users/content seekers to learn to relate older and newer versions.
- To manage accessibility is also a challenge. In information-centric network, the standard assumption is to support ubiquitous access to named data objects, but there are related-use cases in which access to data objects should have restrictions like to a particular user base only. There are distinct ways for this, namely object encryption (needs distribution of key and related methods) or the concept related to scopes, which means depending on identifiers that can only be resolved/used concerning certain conditions.

- To adapt information-centric network as a current Internet architecture, the above-mentioned issues specifically related to naming mechanism in ICN need to be resolved as it is an essential and integral part of its paradigm.

3.5   Naming Approaches Proposed So Far

Certainly, the naming of content lies at the core or in the base for any information-centric network paradigm. Different ICN projects have given distinct approaches for naming content in the network. In this section, a focused and detailed concept exploration of a naming method for major ICN research projects is depicted. The concept of information-centric networks came into the picture in the year 2000, when concept of name-driven routing was introduced by Cheriton et al. in TRIAD [10]. After that, numerous research projects have started working on ICN. In this section, comparison and analysis are done in contrast to various major projects of ICN, mentioned below. These projects give reasonable coverage for the various research-related efforts in the field of naming methods in ICN [1].

- NDN – Named data networking
- DONA – Data-oriented network architecture
- CBCB – Combined broadcast and content based
- NetInf – Network of information
- PURSUIT – Publish Subscribe Internet Technology

In the following sections, a brief description is presented about the naming mechanism adopted in each of the aforementioned ICN projects including the concept, strength, weakness, and important diagrammatic representation.

3.5.1 Combined Broadcast and Content-Based (CBCB) Routing

Combined broadcast and content-based routing is proposed in [11]. It is an overlay at the application level that superimposes a data-driven communication facility over a general network communication called point to point. It adopts a publish–subscribe paradigm, in which the publisher node publishes its data with the help of messages, and the subscriber node advertises its interests with the help of predicates. A message contains a different set of pairs related to attribute values. The predicate is a kind of disjunction of different conjunctions of conditions on specific attributes. The propagation of published messages is done from their source nodes over a broadcast tree. The nodes do pruning of branches of the broadcast tree by using predicates to ensure that message gets delivered to only intended nodes/users.

Naming in CBCB To give name to data inside the network, this scheme takes the help of a set of attribute and value pairs. Each attribute related to data has its name, its type, and a set of different possible values. Let's take an example that demonstrates the CBCB naming mechanism. The name of the data situated at MarwadiUniversity.in/KSdelvadia/ICN_naming.pdf follows the naming convention as mentioned in the Fig. 3.2.

The naming method adopted by CBCB is unique as it is different from the conventional URL-oriented hierarchical naming method and flat naming method used

Fig. 3.2 Naming convention for CBCB

Type-of-File <String>: pdf
Subject <String>: ICN Naming
Author <ListofString>: KSdelvadia
Institute <String>: MarwadiUniversity
Year<Integer>: 2013

in other data-centric network paradigms. But this naming mechanism does not guarantee the uniqueness of names. It does not support even security on names of data.

3.5.2 Data-Oriented Network Architecture (DONA)

The data-oriented network architecture proposes the new approach for naming which is complete redesigning of the existing name resolution service of the Internet. To achieve three major goals mentioned below, it proposes a naming scheme to use a self-certified and flat naming method with a hierarchically structured name resolution system [12].

- Content available in the context of reliability and lower delay
- Persistence of name
- Provenance of data

Naming in DONA Each data chunk in domain-oriented network paradigm is related to a corresponding publishing entity named a principal or owner. So, names of content in DONA follow the format P:L. Here, P denotes a cryptographic hash value for the related principle's public key and L denotes a label that is assigned by principle. The responsibility for granularity and uniqueness of label L must be handled by principle. The names of content are persistent as well as globally unique. The naming methodology is not bounded for any institutional boundaries. Metadata related to each data has a complete public key and the signed digital digest by the principle. The P portion of the name guarantees provenance, and the signature portion inside metadata guarantees the integrity of data. This scheme provides new research opportunities in the field of network caching. Any network node that has valid replica of content can serve the data requests as a content source. Content distribution networks can utilize this function to give access to data from distinct principles, where users receiving those data only require trusting upon content distribution network. The receiver node does the verification process of whether the public key belongs to the principle. The mentioned problem can be solved by two different approaches.

- Store the principal's public key under a specific label that is well known for all.
- Do verification of key based on public key infrastructure (PKI) or web of trust infrastructure (WOTI).

In DONA, the resolution of names is performed by an entity called resolution handlers (RHs), and each RH holds the registration table, which stores 3-tuple data format like <P:L, name of next hop resolution handler, distance>. So the naming of content follows the format <P:L> as already explained above with the significance of each field. For example, any content chunk c1 can have the following name,

<P1:L1>

With P1 as related cryptographic hash value for the principle's public key and principle assigned label L1 to content C1.

3.5.3 Network of Information (NetInf)

Network of information paradigm is subset of the EU FP7 projects named SAIL and 4WARD. The 4WARD project emphasizes more on the naming of the content and searching for the same. The SAIL project emphasizes more on network transport problems. A network of information contributes a naming scheme by using self-certified and flat names same as data-oriented network paradigm [13].

Naming in NetInf Like data-oriented network architecture, network of information names contains two portions namely P:L; here P shows the hash value of the principle's public key and L shows the label selected by the principle. In the case of static data, L represents the hash value of the data itself. But in case of dynamic data, a fixed-length identifier is used as a label L and in order to guarantee the integrity of data, digital signature is used which is stored inside metadata. Network of information provides binding with the help of pair of private/public key to the data in place of the principle; therefore, one principle may use more than one pair of private/public keys. The identification and authentication of principle is found out from the public key chaining data stored inside metadata field. This functionality supports anonymous still secure publishing of data.

3.5.4 Named Data Networking (NDN)

The named data networking is a part of the four projects of National Science Foundation's Future Internet Architectures (NSF FIA) in order to explore, design, and build the future paradigms for Internet. The goal of the named data networking is to redesign and rebuild the entire Internet paradigm from scratch. It actually replaces IP with data chunks as a global component for transport [14].

Naming in NDN The names in named data networking are made up of more than one component organized hierarchically as demonstrated in the figure. The component is a kind of string having an arbitrary length. The transport layer of named data networking provides no restriction for names/identifiers except the structure of the component. Names are getting produced and assigned by the end users. Named data networking only gives the structure and expects that norms related to naming come out and become standardized by the development of distinct categories of applications. Content names have data-like segment numbers and version. All content names inside named data networking by default cover a SHA256 digest of the data for resolving the error. To assure integrity and authenticity of data, the mappings of data to name are signed digitally and sent with the data. Because of the hierarchical name structure, named data networking names are nonpersistent and human friendly in nature. The end-user assigned portion of a name may not have unique nature but the complete name of a data along SHA256 digest value is unique. The hierarchical name structure of named data networking paradigm of future Internet has been demonstrated in the following Fig. 3.3[14].

3.5.5   Publish Subscribe Internet Technologies (PURSUIT)

Publish Subscribe Internet Technologies is a project of EU FP7 that has been recently released as a follow-up to a past project of EU FP7 named Publish–Subscribe Internet Routing Paradigm (PSIRP) [15]. PSIRP has proposed a routing paradigm that is clean slate for information-centric networks that shifts the existing send–receive-driven Internet to the publish–subscribe-based architecture. The major objective of the PURSUIT project is to develop a deployable Internet-scale component for the publish–subscribe Internet-routing paradigm.

Naming in PURSUIT Publish–Subscribe Internet-Routing Paradigm adopts the same naming mechanism as data-oriented network architecture. The data names are termed Resource identifiers (RIds). The same naming method is used by PURSUIT project [16]. Persistence of data is guaranteed with the use of specific network elements named content sources that reside at the network edge. There is the periodic refreshment of states of data publication inside the network by content sources.

The publish–subscribe Internet routing protocol network is dependent on the concept called scopes; here scopes are recognized with the help of Scope identifiers (SIds). The authorization, availability, reachability, persistence, replication, access

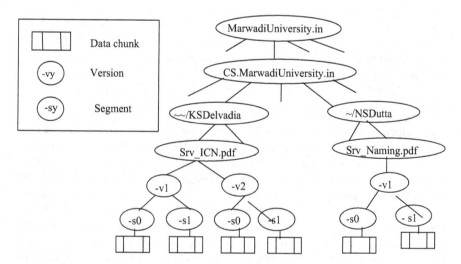

Fig. 3.3 Hierarchical name structure in NDN

rights, and upstream resources of data are controlled by scopes. Data request (subscribe) and data publication (publish) are dependent on the pair of <Sid, Rid>. A default assumption for this method is that data publishers publish and data subscribers subscribe for data within a scope that they do trust.

3.5.6 *Hierarchical and Flat-Based Hybrid Naming Scheme in CCN IoT*

The work presented in [9] proposes a hybrid naming method for the content names in CCN-based Internet of things (IoT) scenario. Internet of things needs to push as well as pull both communication modes along with secure and scalable data names in context of data integrity. In this research work, they have proposed a hybrid naming mechanism that assigns names to data with the help of flat and hierarchical components in order to support pull-and-push communication and to assure security and scalability. Any CCN-oriented IoT data naming method requires support for location independence, global unique retrieval of data, push support, security, and readability for humans.

The proposed hybrid naming mechanism combines positive aspects of flat as well as hierarchical naming methods and gives a hybrid naming method that contains two major portions namely, (1) flat component and (2) hierarchical component. With the help of simulation, they have proved that the proposed method improves the interest rate transmissions, aggregation of name, number of covered hops, and reliability [9]. It also addresses the loop problem related to communication. A thorough description of both of the components of the proposed schemes is found here.

1. *Hierarchical component:* This module is produced by following uniform resource identifier (URI) syntax that covers three subparts as mentioned below.

   (a) Name of the domain
   (b) Location
   (c) Task

here, the forward slash "/" is used to provide separation among subparts of the hierarchical component. These three subparts are merged to form the complete hierarchical part. Like,

3.5.6.1 DomainName/Location/Function

The above syntax can have values like MarwadiUni/CA/DC/action: off as demonstrated in Fig. 3.4.

The definitions of these subparts are presented in following Table 3.1. The hierarchical portion works as a prefix to (1) locate domain name, (2) data retrieval site,

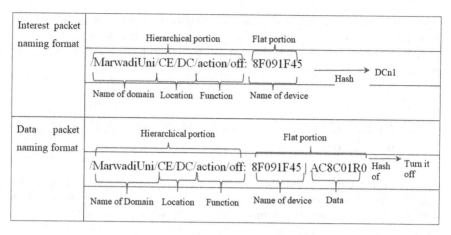

Fig. 3.4 Naming format for interest and data packet

Table 3.1 Definitions for subcomponents in naming format

Subcomponents	Significance
Name of domain	It is useful to represent institute name
Location	It is useful to show the node location where interest packet is sent and from where to receive the content
Function	It is useful to show the functions that can be executed like action and sensing
Name of device	Flat portion that is useful to show device uniquely by performing encryption on a name
Content	The flat portion does the encryption of content to assure the integrity of it

and (3) location of the function to be executed. The function portion specifies the function types like sensing and action. Aggregation of names can be done with the help of this representation in order to achieve routing table optimization.

2. *Hash/flat portion:*

The above hierarchical portion is merged with another flat portion with the help of ":" symbol as a separator. The flat portion deals with the content and name of the device. This portion has fixed 32-bit length so that problems related to lengthy hierarchical names can be avoided. It assures uniqueness in names though it has low support for aggregation. To provide security on content produced and on-device, a non-cryptographic hashing method named FNV-1a hash [17] is used. Let's take an example, consider the computer engineering department of Marwadi University, node n1 in the lab of data computing, then algorithm works as mentioned below.

h=offset factor (for every octet chunk to be hashed)
h=h (xor) octet chunk
h=prime factor*h
Return the h

Here, h shows the value of hash and octet shows the data octet/chunk. The value of the prime factor and offset factor depends on the hash length. The values of prime factor, offset factor for x-bit hash and hash for DCn1 are referred from [18]. The same values you can correlate inside Fig. 3.4. In addition to that, the symbol "|" is used to do a separation between interest packet data name and data packet content.

3.5.7 Secure Naming Scheme for Information-Centric Networks

The proposed naming scheme in [5] presents a secure naming method to identify/ locate different resources inside ICN. The major objective is to permit the secure retrieval of data from the different untrusted or unknown sources inside the network. The proposed scheme is flexible in the sense that it has backward compatibility with the existing URL-based naming method. It also permits independent data recognition irrespective of the forwarding, routing, and storage methods by isolating the location recognition rules and a source inside URI/URL authority components. With the help of the proposed naming scheme, you can securely receive data from any of the sources inside a network. It also supports data mobility, data validation for original/actual source, and complete backward compatibility with the existing naming method.

The implementation of the proposed approach needs two major components: (1) content manager and (2) name resolver. Name resolver can be deployed as a plug-in inside web browser to support data retrieval from different sources. For any given URL, this plug-in first checks whether it contains the certificates inside its trusted store of a key in order to authenticate the authority. Else the plug-in starts a name resolution method to receive the certificate of authority along with the data identifier, carries out the authentication, and stores it inside the local store of the keychain. Then, the name resolver sends a request to the domain name system to map authority name to metadata and the network location of the webserver. Figure 3.5 shows the header format used in the data chunk retrieval.

Here, the type of field specifies the message type. The second field specifies the context of the needed content or ID of the data provider. The third field is the cryptographic identifier of the data, to protect against unauthorized changes inside data. The fourth field carries the length of the actual data that present inside the fifth field.

Type	Context/ID of authority	Piece crypto ID	Length of piece	content

Fig. 3.5 Data piece header

3.6   Analysis of Naming Schemes Along with Their Advantages and Disadvantages

In this section, a comparative analysis of the existing naming mechanisms proposed so far is carried out for several important factors. This comparative analysis will help readers to have some insight to propose the new efficient naming mechanism for information-centric networks.

3.6.1   Flat vs. Hierarchical vs. Attribute Value-Based Naming Approaches

The existing naming-related schemes can be categorized in below three major categories: flat, hierarchical, and attribute–value pair oriented [1]. All three types of names can be aggregated to some extent to improve the scalability of the routing table. Prefix aggregation for hierarchical names is done by NDN to support scalability. CBCB excludes the not-needed branches of a broadcast tree with the help of attribute–value pairs. Predicates that have same attributes at the router can be merged so that the number of entries in routing table can get reduced in CBCB. DONA, MDHT/NetInf, and PURSUIT use the flat names which are in the form P:L and at the publisher level, they can be aggregated. But in the case of DONA, this aggregation is not the efficient one. That is the reason that DONA has the problem of heavy load at upper-level RHs. Flat names are preferable for DHT-oriented lookup functions like PURSUIT and NetInf, where there is an equal distribution of storage load among the resolution nodes. PURSUIT also does the other level of aggregation for names at the scope level (sID).

Human-Friendly vs. Flat Names The involvement of cryptographic hash value inside data name hides the inner meaning of data from human users. This also creates difficulty in remembering the names. Because of this, self-certified flat names used in NetInf, DONA, and PURSUIT are not human friendly. In contrast to that, the naming methods used in NDN and CBCB are human friendly due to their attribute-oriented partition or hierarchical structure. This creates easiness in remembering names as well. And names also can provide more semantic-oriented information of content. But to achieve these benefits, there are some challenges as well like security binding, authenticity, and assuring global uniqueness. This discussion leads to one important basic aspect like the necessity of the naming system as human friendly. Also with the possible vulnerabilities related to security and the self-certified names as well as flat and location independence in nature to guarantee the integrity of data [1].

Names that are not human friendly are not a hurdle while accessing data as long as the data contains specific keywords involved in it. The user is concerned about receiving the authentic data. So it may be concluded that in the context of ICN, names of data should be less human friendly to cover properties like the capability of self-certification and location independence.

3.6.2   Name-Based Routing vs. Name Resolution

The majority of information-centric networking-naming mechanisms are dependent on the concept related to identifier/locator split. All the existing methods for finding data based on its identifier which is location independent can be divided under two main categories called resolution of name and name-driven routing [1]. The name resolution process contains two major phases: (1) The name of the data is getting resolved into one or more IP addresses or locators. (2) The content request is routed toward one of these IP addresses with the help of any of the shortest path routing protocols like OSPF or ISIS. In contrast to that, in name-driven routing, the forwarding of request is done depending on the name directly. The network also maintains the state information along the route so that the data can send back to the user/requester. So out of all the research projects proposed above, PURSUIT and NetInf had adapted the name resolution concept and CBCB, NDN, and DONA had adapted the name-driven routing concept.

Though DONA, CBCB, and NDN had adapted name-driven routing concept, their methods for naming are distinct. NDN supports hierarchically structured names, DONA supports flat names and CBCB supports a set of pairs of attribute–value to name a data. DONA has a hierarchical infrastructure for the resolution of the name. The approach of NDN for routing is same as existing IP-based routing. But there are names of data in place of IP addresses. Still, CBCB is special as its working methodology for routing is a mixture of publish/subscribe-oriented routing and flooding.

The concept of name resolution can assure the finding of any data inside network for fixed network size. But name the driven routing concept does not assure finding of data. In place of that, they can assure a high likelihood of data discovery, which is in a major case, proportional to the count of visited network nodes. Failure of nodes can create trouble in the name resolution system though data is present the same problem does not exist in name driven routing system. In addition to that storage needs for the name resolution approach is more than the name driven routing approach. There are two different databases maintained by name resolution approaches like name of the content and IP address mapping inside resolution system and reachability information of IP addresses inside routing system. But in the case of name-driven routing concept, there is only one database that contains the mapping between names of data and network addresses.

3.6.3 Incremental vs. Clean Slate Approach

All the research projects of ICN can be divided into two types, like deployment of some projects need clean slate while other deployments can be incremental or coexist along with existing technologies. NDN, CBCB, and PURSUIT have a clean-slate design that does not need IP-based routing. But NetInf and DONA give a new name resolution system that replaces DNS, while still make use of existing IP-based routing protocols like OSPF, RIP, BGP, etc. So both DONA and NetInf can be incrementally deployed, one Internet service provider at a time. The attribute called incremental deployment is more preferable if you consider any practical aspect of the solution; hence NetInf and DONA are more preferable over others in the said context [1].

Scalability As per the analysis report of BGP routing table, the biggest routing table on the Internet has approximately 4×10^5 routes of BGP to cover approximately 3.8×10^9 IPv4 addresses and about 6×10^8 Internet hosts [1]. This scaling factor of 10^4 between BGP paths and IPv4 an address is achievable by name-based routing and aggregation of route. Still, count of addressable contents inside ICN is predicted to be of several more orders of magnitude. As per the current scenario, there are approximately 10^8 second-level domains which are registered inside Internet. This figure is the same as number of data publishers in ICN. Though aggregation can be performed on names at publisher level, the name-driven routing table require to manage approximately 10^8 paths, which is much more greater than the largest routing table size of BGP. So, all the ICN research projects discussed so far have faced the problems related to scalability in the context of Internet-wide deployment. Table 3.2 shows the comparative analysis of different ICN projects in context of naming and several affected performance parameters.

3.7 Need of an Efficient Content Naming Mechanism: Our Perspective

Based on that analysis carried out so far, the perspective about the selection of an efficient naming method for ICN is figured out under the following parameters.

Structure of Name It is believed that in place of using human-friendly names for web data, it will be efficient if that content contains a set of owner-chosen keywords that have been assigned to it. Later on, the search engines can use these keywords to index them and also allow users to do searching for the needed data.

Self-certification To name data, flat names that are self-certified can be used. The major two reasons behind it are as follows:

Table 3.2 Comparative analysis of existing naming schemes

Naming scheme proposed in	Approach	Human friendliness	Name resolution or name-based routing approach	Clean slate vs. incremental approach	Scalability
CBCB	Attribute–value-based naming	More	Name-based routing	Clean slate	Less
DONA	Flat naming	Less	Name-based routing	Incremental	Less
NetInf	Flat naming	Less	Name resolution	Incremental	Less
NDN	Hierarchical naming	More	Name-based routing	Clean slate	High
PURSUIT	Flat naming	Less	Name resolution	Clean slate	Less
Hybrid naming for CCN-based IoT	Flat + Hierarchical naming	Average	Name-based routing	Clean slate	High
Secure naming in ICN	Hierarchical naming	Average	Name resolution	Clean slate	High

(a) This approach provides persistence of name, security binding, global uniqueness, and authenticity.

(b) The analysis based on scalability given by DONA as mentioned in [18] recommends that the flat naming–based routing approach is more suitable for the technology of today.

So data in ICN should be globally unique, independent of location, secure, and should have user-friendly names. In reality it is hard to find the single naming method that fulfills all the above functionalities. Instead of it, a multilayer naming mechanism that has self-certified names along keywords that are human friendly is the best match in practical applications. Hierarchical naming gives hierarchical human-readable names. The benefit is its compatibility with existing system and data aggregation to reduce the routing data while increasing the efficiency for performing search inside routing table. The drawback is the name of data which has variable length and long in size. The flat naming scheme guarantees independence of location and application as well as global uniqueness. Still, it slows down the procedure of aggregation which increases the routing table size. The flat names sometimes also become unreadable.

Because of this, it is advisable to use the hybrid approach for efficient naming mechanism in ICN as it combines the advantages of both hierarchical as well as flat naming scheme. It is also suitable for ICN deployment for the existing Internet paradigm. One such hybrid scheme is explained in the literature survey. It can get extended or modified to make it more applicable and efficient for current Internet user needs.

3.8 Open Research in ICN Naming

An efficient naming strategy plays an important role for the successful ICN deployment. The research community across the globe is investing their efforts in order to explore this research area since the last decade. The selection of naming mechanisms influences the overall performance of an ICN network in context of routing, caching, and security as well. Therefore, it's a crucial and challenging task to select an appropriate naming mechanism for ICN deployment. Several important factors and open research directions in context of ICN naming have been explored herewith.

There are majorly two categories for naming mechanisms in ICN, namely flat and hierarchical. Flat identifiers are formulated by applying hashing algorithms to existing data chunks. It is not suitable for dynamic content that is not published yet like in the case of IoT. In opposition to that, hierarchical identifiers support the naming of dynamic content that is being generated based on demand, given that the naming scheme has been mentioned at the time of system setup. Still, hierarchical identifiers are influenced by constraints on length for example to accommodate with highest payload size for protocols like ZigBee. Apart from this, variable-length identifiers lead prefix lookup at line speed, a very challenging task. Specifically for large-scale network scenarios, naming methods must be designed along with comparative analysis of processing mechanisms like encoding of prefix components. This accelerates the name prefix lookup from FIB. This is the core task to minimize data retrieval delay. This plays a very crucial and important role in the field of smart healthcare and intelligent transport systems.

The selection of the naming method is a core criterion in ICN that determines the operational separation among the application layer and network layer. The selection of naming schemes influences the scalability and performance of ICN in the context of routing efficiency, FIB size, and the size of a namespace. The effectiveness of network paradigms can be measured with the help of flat and hierarchical identifiers through distinct metrics like forwarding efficiency, semantics, and aggregation ability. Though these metrics are independent in nature, forwarding efficiency can get affected if there is lack of aggregation ability due to large FIB tables. Replication and content mobility inside the network can influence the aggregation ability and raise FIB size in case of hierarchical naming as well. The communication pattern and semantics can influence management of namespace and lead to distinct table sizes.

The authors in [6, 13] have done a comparative analysis between distinct naming schemes and their impact on performance metrics of network along with related benefits and limitations specific to network applications. Though the choice for naming scheme is specific to the performance requirement for a particular network application, it is always desirable to do a deeper study while selecting anyone as this is a key driving force behind ICN performance at the user as well as network level.

3.9 Summary

In this chapter, a thorough discussion on the basic concepts related to naming in information-centric networks. The significance of the same and existing naming mechanisms used by several ICN research projects with related standard naming format along with its related strengths and weaknesses is included in the discussion. A comparative analysis of such schemes concerning several competitive design factors and insight toward an efficient naming solution in ICN. The chapter may be concluded by analyzing that there is a clear lack of the efficient and ideal naming method for ICN and this gap gives a starting point to all the researchers that are keen to work in this exciting field of research.

References

1. Bari, M., Chowdhury, S., Ahmed, R., Boutaba, R., Mathieu, B.: A survey of naming and routing in information-centric networks. IEEE Commun. Mag. **50**(12), 44–53 (2012)
2. Walfish, M., Balakrishnan, H., Shenker, S.: Untangling the Web from DNS. In: Proceeding of the 1st Conference on Symposium Networked Systems Design and Implementation — Volume 1, ser. NSDI'04, Berkeley, CA, 2004, pp. 17–17.
3. Pan, J., Paul, S., Jain, R.: A survey of the research on future internet architectures. IEEE Commun. Mag. **49**(7), 26–36 (2011)
4. Zhang, M., Luo, H., Zhang, H.: A survey of caching mechanisms in information-centric networking. IEEE Commun. Surveys Tuts. **17**(3), 1473–1499., 3rd Quart (2015)
5. Wong, W., Nikander, P.: Secure naming in information-centric networks. In: Proceedings of the Re-Architecting the Internet Workshop, (ReARCH'10). pp. 1–12 (2010)
6. Ahmed, S.H., Bouk, S.H., Kim, D.: Content-Centric Networks An Overview, Applications and Research Challenges, Content-Centric Networks, 2016 ISBN: 978-981-10-0064-5.
7. Yu, K., Arifuzzaman, M., Wen, Z., Zhang, D., Sato, T.: A key management scheme for secure communications of information centric advanced metering infrastructure in smart grid. IEEE Trans. Instrum. Meas. **64**(8), 2072–2085 (2015)
8. Adhatarao, S., Chen, J., Arumaithurai, M., Fu, X., Ramakrishnan, K.: Comparison of naming schema in ICN. In: IEEE Local and Metropolitan Area Networks, June 2016.
9. Arshad, S., Shahzaad, B., Azam, M.A., Loo, J., Ahmed, S.H., Aslam, S.: Hierarchical and flat-based hybrid naming scheme in content-centric networks of things. IEEE Internet Things J. **5**(2), 1070–1080 (2018)
10. Cheriton, D., Gritter, M.: TRIAD: a scalable deployable NAT-based internet architecture. Technical Report, Jan. 2000, available: http://www.dsg.stanford.edu/triad/
11. Carzaniga, A., Rutherford, M., Wolf, A.: A routing scheme for content-based networking. In: INFOCOM 2004, 3rd Annual Joint Conf. IEEE Computer and Commun. Societies, vol. 2, Mar, 2004,vol. 2, pp. 918–28.
12. Koponen, T., et al.: A data-oriented (and beyond) network architecture. SIGCOMM Comput. Commun. Rev. **37**, 181–192 (2007)
13. Dannewitz, C., et al.: Secure naming for a network of information, In: INFOCOM IEEE Conf. Comp. Commun. Wksp., 2010 2010, pp. 1 –6.
14. Jacobson, V., et al.: Networking named content. CoNEXT, J. Liebeherr et al., Eds. ACM, 2009, pp. 1–12.
15. Lagutin, D., Visala, K., Tarkoma, S.: Publish/subscribe for internet: PSIRP perspective. In: Towards the Future Internet Emerging Trends from European Research, vol. 4, pp. 75–84 (2010)

16. Fotiou, N., et al.: Developing Information Networking Further: From PSIRP to PURSUIT. In: Int'l. ICST Conf. Broadband Communications, Networks, and Systems (BROADNETS), 2010 (invited paper), Oct. 2010.
17. Fowler, G., Noll, L. C., Vo, K.-P., Eastlake, D.: The fnv noncryptographic hash algorithm, Ietf-draft, 2011.
18. Fowler noll vo hash function. [Online]. Available: http://will.thimbleby.net/algorithms/doku.php?id=fowler-noll-vohashfunction

Chapter 4
Routing Schemes used in Content Delivery

4.1 Introduction

Data transmission in ICN is initiated by the consumer of data through the generation of an interest packet. The interest packet carries the unique name of the content requested. When an interest packet reaches a content publisher or a node having valid requested content, a data packet is retraced to the content requestor in the reverse path of Interest. The data packet also carries the unique name of the content. The node that sends a copy cached in its store is called "provider," and the node which generated or originated the data is termed as "producer" [1]. These two terms are used throughout this chapter to distinguish between the data-caching routers and the actual producer.

The routing schemes in ICN deal with the preparation of network topology and strategies for handling long-term changes in the network. They are also responsible for preparing and updating the forwarding table. The underlying routing methods coordinate with ICN's forwarding module for identifying the incoming and outgoing network interfaces to decide for data transmission [2, 3]. Unlike IP networks, the ICN has difference between routing and forwarding. In ICN, routing provides information about the availability of a path (or route) to content, whereas the forwarding selects the appropriate route to the location of the content based on performance. Although IP-based routing protocols like OSPF, link–state, etc. are stable enough, they cannot be used directly with ICN. Because these algorithms are host centric and ICN is content centric. However, with some modifications they good enough to use in ICN [4].

© Springer Nature Switzerland AG 2021
N. Dutta et al., *Information Centric Networks (ICN)*, Practical Networking,
https://doi.org/10.1007/978-3-030-46736-4_4

4.2   Realization of Routing in ICN

A content router maintains a Pending Interest Table (PIT), a Forwarding Information Base (FIB), and a Content Store (CS). The FIB, which stores the reachability information of content names, is used to forward interest packets toward the content source. The PIT of a content router records the interface that interest packets arrive at the content router so that the returned data packets can be correctly sent to their original requester. Note that if a content router receives interests for the same content chunk from multiple interfaces, it should record all these interfaces in its PIT. The CS is used to cache some recently forwarded content chunks. When an ICN content router receives an interest packet for a content chunk named by a unique content name, it directly responds to the interest packet with the corresponding data packet if it caches the copy in its CS. Otherwise it inquires its PIT. If the content router finds an entry for the content name in its PIT, it records the interface from which the interest packet arrived at the content router. If the content router cannot find an entry for the content name in its PIT, it forwards the interest packet toward the content source after querying its FIB. When a content router receives a data packet, it firstly caches the Data packet into its CS. It then queries its PIT and forwards the data packet to the next hop by sending out the Data packet to the interface recorded in the PIT entry for the corresponding content name. After that, the content router deletes the entry for the content name from its PIT. Figure 4.1 shows a content router with its various data structures.

The task of any of the routing protocols in ICN is to prepare the FIB table by inserting a path to the named content as it will be referred to forward interest packet toward the content source based on the routing paths mentioned in this table. FIB carries reachability information associated with each name prefix. So, a router can get the next hop to visit and reach that content source from FIB. One FIB entry is created for each next hop for a name prefix. Different routing mechanisms propose distinct solutions and strategies to fill the FIB table. The majority of research work has extended the existing routing protocol like OSPF and LSR for NDN architecture. However, few other innovative protocols are also proposed by various researchers. In this section, a discussion about such routing protocols is carried out.

Fig. 4.1 ICN router as defined by NDN architecture

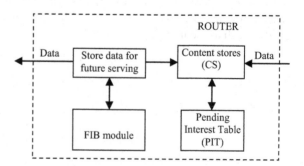

4.3 Realization of Forwarding in ICN

The NDN-forwarding module is designed to detect various failures like node, link, or packet. It also accomplishes the task of recoveries from such failures. It has the advantage that the routing module need not have to perform a lookup in a continuous FIB table, thereby improving the scalability and stability of routing [5, 6]. The forwarding strategy has two functionalities: scalability assurance and forwarding. Scalability provides the means of measuring packet's RTT, throughput, and packet loss statistics. It also finds an alternative path during congestion.

Providing scalability is a challenge in NDN because of the use of unique but unbounded variable-length content naming. To provide scalability, there is a need for a fast low-cost name lookup method. The NDN architecture also has a read–write forwarding plane, which demands per-packet information updates at underlying line speed. Forwarding is stateful as NDN content router maintains a PIT table to keep track of all incoming and not-replied-to interest packets. The PIT entries comprise a tuple having five fields like content_name, nonce, incoming and outgoing interface(s), and timer.

4.4 NDN Forwarding Architecture with Illustration

As per the discussion in the previous section, the data retrieval process in NDN depends on two types of packets, namely data packet and interest packet. When a user needs certain data, it sends an interest packet to the network, which can be answered with the related data packet from the network. To realize this, NDN content routers disseminate reachability data of identifiers with the help of routing protocols like OSPF and BGP similar to the current Internet which disseminates the reachability data of IP prefixes. Also, the content router has three data structures, namely Pending Interest Table (PIT), Forwarding Information Base (FIB), and Content Store (CS). The FIB table stores the reachability data of identifiers which will be referred to forward interest packets toward the data source. The PIT table will store the information about the router interface at which the interest packet arrives. This can be helpful to send back the data packet to the original requestor. If a content router gets an interest packet for the same data chunk from distinct interfaces, a PIT will record all these interface details. The CS caches some of the recently forwarded data chunks.

When content router in NDN gets an interest packet for a data chunk entitled by a unique data identifier, it responds immediately with the related data packet if it has already cached the copy of needed data inside its CS. Else it queries its PIT. If the corresponding data identifier entry is found in its PIT, it will store the interface data from where the interest packet reaches the content router. But when the content router is not able to find the related data identifier entry inside PIT, it sends the interest packet toward the data source after inquiring about its FIB table.

When the content router in NDN retrieves a data packet, first it caches the data packet inside its CS. Then it inquires its PIT and sends the data packet to the next node by referring to the interface details stored inside the PIT table for related data identifier. Afterward, the content router discards the PIT entry related to that data identifier. Figure 4.2 is an illustration to understand NDN forwarding paradigm. Assumed that data source contains the needed data named EX. And the reachability information for this content identifier is disseminated with the help of routing protocols like OSPF or BGP. Therefore, FIB tables of CR1 and CR2 contain a record for that content identifier as shown in (1) and (4) in Figure, respectively. To get the needed data EX, user A will first send the interest packet containing data identifier EX to its 1-hop content router named CR1. CR1 cannot find the relevant cached data inside its CS, so it looks inside its PIT table, which also doesn't have the related data identifier entry. So it will create an entry for that identifier in its PIT table as depicted in (2) of Fig. 5.2. Afterward, CR1 inquires its FIB table and sends the interest packet to CR2 through interface if2. Similarly, when CR2 gets the interest packet, it cannot find the related cached content inside CS and related content name entry in its PIT. So CR2 also creates a PIT entry with content name EX and interface details as shown in (5) of Fig. 5.2. Then CR2 will inquire about its FIB table and sends an interest packet toward the data source.

When the data source retrieves the interest packet, it sends the needed data packet to CR2. When CR2 gets the data packet, first it stores it inside its CS. After that, it queries its PIT table and came to know that it should forward the data packet through interface if4 and it sends data packet through if4 and removes PIT entry for that content name related to data packet as shown in (6) of Fig. 5.2. In the same manner when CR1 gets the data packet, first it stores it inside its CS and forwards the data packet through interface if1 and removes PIT entry for that content name as shown in (3) of Fig. 4.2. Now let's say user C connected to CR2 by interface if6 also needs

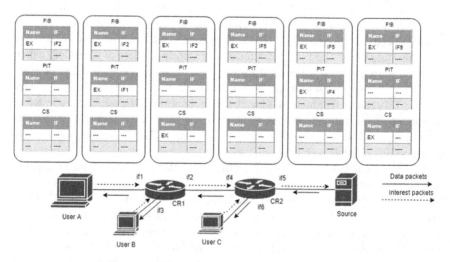

Fig. 4.2 NDN forwarding architecture

to fetch the data named EX. For this, user C forwards an interesting message to CR2. Now when CR2 gets the interest message, it can successfully search a cached copy of the needed data inside its CS. So, CR2 immediately forwards the needed data message to the requestor, without getting it from the data source. So, it is seen how the content retrieval and forwarding occur in NDN with illustration and supported fundamental data structures. It summarizes the functional architecture for the content forwarding in NDN. An in-depth discussion on the stateful forwarding paradigm of NDN is coming up in the next section.

4.5   Stateful Forwarding Paradigm for NDN

A network's paradigm design decides the flow and form of the related forwarding mechanism. The current IP-based network achieves delivery of packet in two distinct phases. At the level of the routing plane, routers interchange routing updates and choose the best paths to build a forwarding table. At the level of the forwarding plane, routers use to route the packets by referring to the forwarding table. Therefore, IP-based routing is considered adaptive and stateful, while IP-based forwarding is considered stateless and inflexible. This dumb forwarding and smart routing mechanism assign the responsibility of secure content delivery entirely to the routing system. As a result, IP's routing-level plane is also called the control plane, and its forwarding-level plane is the data plane. Routing in NDN serves the same motive as in the case of IP network which means calculating routing tables that are to be referred to in process of NDN's interest message forwarding. The NDN's forwarding-level plane is divided into two steps: first requestors send out interest messages, then after data messages travel back following the same route in the reverse way. Content routers maintain state information related to pending interests to guide data messages back to original requestors. The advantages of NDN's forwarding level plane cover in-built network caching, multicast content delivery, and adaptive forwarding. By storing pending interests and monitoring data messages traveling back, each content router can compute message delivery performance like throughput and round-trip time (RTT). It can also detect troubles that can cause packet losses like congestion or failures of links. By using this adaptive and intelligent forwarding plane, the NDN's routing plane only requires to propagate long-term updates in policy and topology without dealing with short-term churns.

The stateful forwarding paradigm of NDN in terms of interest and data packet processing is illustrated in Fig. 4.3. This figure is a graphical resemblance of the stateful forwarding process described in Sect. 4.4. The NDN content router maintains a PIT entry for each pending interest message; it can be said that NDN router preserves a "datagram state." This state causes a closed-loop and two-way symmetric message flow: on each communication link, each interest message pulls back only one data message, preserving one-to-one message flow balance, excluding rare incidents when messages get lost in-network or required content does not exist in actuality. The datagram state of NDN is different from the virtual circuit state of

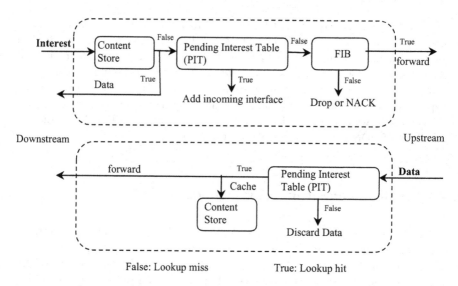

Fig. 4.3 NDN's stateful forwarding paradigm

MPLS or ATM. Firstly, a virtual circuit establishes a single route among ingress–egress router pairs and when it demolishes, the state needs to be recovered for the whole route. Secondly, a virtual circuit describes the route to be utilized to forward packets; due to congestion if any of the communication links along the route gets overloaded, messages on the same virtual circuit cannot be redirected for adaptability of changes in load. In contrast to that, the datagram state of NDN is per interest, per node. At each node, the router takes decisions related to where to forward the interest packet on its own. In case of router crash or link failure, it only affects the interest packets at that particular location; the routers present at previous-hop can recognize the failure quickly and deal with the problem-creating areas. In this section, exploration is carried out on the basic design on how to use the datagram state of NDN's router to build an adaptive and intelligent forwarding plane. The major objective was to fetch content through the best performing route(s) and to detect any message delivery problem quickly and to get recover from it. Figure 4.3 introduces a packet named Interest NACK. Whenever a NDN node called N_u can neither fulfill nor further forward an interest packet, it will send a packet called Interest NACK back toward the downstream node named N_d. Now if node N_d has finished all of its forwarding alternatives, then it will send a NACK packet toward further downstream nodes. One key thing to remember is that while interest messages flow from downstream node to upstream node, data messages flow from the upstream node to downstream node, and Interest NACK messages are always forwarded from upstream nodes to downstream nodes. Sections 4.4 and 4.5 explore in depth how NDN's forwarding architecture works and the significance behind NDN's stateful forwarding paradigm, respectively.

4.6   OSPFN: An OSPF-Based Routing for NDN

The first reported NDN routing protocol is Open Shortest Path First (OSPF) for named data or so-called OSPFN [7]. This protocol is an extension of the popular IP-based routing algorithm OSPF. The OSPFN has implemented the concept described by Jacobson, the designer of the named content [8]. So far, the OSPFN has undergone two generations, the version 1.0 and 2.0. The OSPFN version 1.0 has been implemented by simple modifications of IP-based OSPF concept into NDN architecture. However, version 1.0 has been implemented with significant upgradation of name-based routing and multipath routing support. OSPFN comprises three modules to accomplish the task of routing. These modules are OSPFN, OSPF Daemon (OSPFD), and Content-Centric Networking Daemon (CCND). The OSPFN module is responsible for building content names based on OLSA [9]. The constructed object names are then forwarded to the OSPFD to announce the name to the network. On receipt of an OLSA from the other node, the OSPFD forwards the same to the OSPFN of the same node. OSPFN also sends route query to OSPFD for discovering route of each name prefix. CCND keeps track of the FIB entries as described in NDN architecture. OSPFN forwards FIB entries such as name prefix, next-hop(s), and path cost to record the information by the CCND.

In NDN, routers forward interest packets by referring to their FIB using name contained in the packet. Instead of manually configuring FIB, which is error prone and takes too much time, a routing protocol that dynamically produces routes to name prefixes depending on the topology of network is required. To address this challenge, OSPFN is designed as a solution by extending OSPF. As OSPF, the new extended version of the OSPFN also follows a link–state routing approach through a link–state database (LSDB). To build LSDB, each router collects link–state information related to the network. This database is updated based on periodic link–state advertisements. All the routers contain the same LSDB Copy with them and create network topology from LSDB and execute the shortest path first method to compute the route toward each destination node. The routing table is recalculated in case of any change in topology. It also provides support for multipath with an equal cost which means it generates distinct routes to the same destination if they have equal lowest cost across all available routes. Proposed OSPFN protocol supports configured multipath routing approach with name-based routing. To announce name prefixes, they have used OSPF's OLSAs. The open-source routing protocol suits like Quagga OSPF gives an API that permits easy insertion and retrieval for OLSAs through OSPF protocol.

Each NDN router runs three daemons: OSPFN, CCND, and OSPFD. The inter-action among CCND, OSPFN, and OSPFD is illustrated in Fig. 4.4. The significance of each call out message sequence number is specified here. (1) The OSPFN generates a route toward name prefixes. (2) The content-centric network daemon (CCND) is responsible for the management of forwarding of interest and data packets. The OSPFD is responsible for flooding the OLSAs to the whole network. (3) The OSPFN prepare Name OLSAs and (4, 5) include them into local

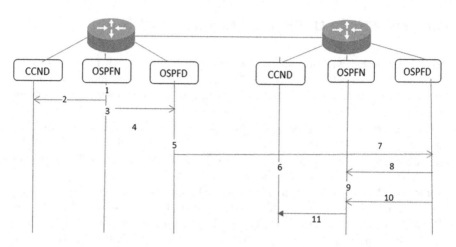

Fig. 4.4 Interaction among CCND, OSPFN, and OSPFD

OSPFD. (6, 7) When OSPFD of a node gets OLSA, (8) it delivers it to its local OSPFN. (9) As each LSA contains ID of the originator router, the OSPFN can retrieve the router ID corresponding to Name OLSA and query OSPFD to get the corresponding next hop to reach the router. (10) OSPFN retrieves the answer of a query from OSPFD which has the next hop value in it. OSPFD continues to flood LSAs and determine the shortest path based on network topology. (11) Then OSPFN installs the name prefix and related next hop inside the FIB of CCND. Each Name OLSA message contains one name prefix along with a set of predefined fields related to flooding scope, name prefix size, and format as well as field related to application-specific use.

4.6.1   Routing in OSPFN

Various message communications for the routing process are shown in Fig. 4.5. The significance of each call out message sequence number is specified here. Whenever a router starts, the booting process (1) reads the configuration file and build name OLSA for individual name prefix. This name OLSA is desired to advertise in the network. Name OLSAs is sent to local OSPFD (2), which in turn get flooded through the network. OSPFD notifies OSPFN in case of any updates in its LSDB along with the content of LSA (3). After getting LSA, OSPFN first verifies whether the received LSA is OLSA or not. If not, then it is discarded by OSPFN. If yes, then the router first verifies whether it is produced by itself. If not, then OLSA is processed (4). OSPFN then gets the name prefix and inserts an entry in its name prefix table, which has a list of name prefixes and related origin routers (5). The OSPFN sends a query message to OSPFD to retrieve the next hop to go to the original router of each name prefix (6). When OSPFD gets a query from OSPFN, it first finds related next hop and associated cost in its routing table, includes this in a single message (7), and sends it back to OSPFN (8). When OSPFN receives a message (9), it updates its name prefix table regarding the next-hop list and route cost for all name prefixes entries that have this particular router as their origin router. Afterward, OSPFN builds FIB entries for individual name prefixes and include those into CCND (10). A single FIB entry is created for each next hop of a name prefix. When OSPFN gets any message from OSPFD regarding deletion of name OLSA, first it deletes corresponding name prefix entries from its name prefix table (11). Then OSPFN notifies CCND to delete relevant FIB entries from CCND (12). The figure shows the sequence of message exchanges among OSPFN, OSPFD, and CCND (Fig. 4.5).

OSPFN does not execute shortest path computation; instead, it sends the query to OSPFD to retrieve the next hop of the origin router for a name prefix. After receiving answer to a query, OSPFN creates an FIB entry that contains the name prefix and related next hop. If multiple routers advertise name prefix, OSPFN sends a query for each and create FIB entries for the name prefix and each next hop. By default, OSPF gives a single next hop for each destination except when it has multiple same-cost shortest routes. OSPFN by default produces a single route for each name prefix. Still, the unique characteristic of NDN is its multipath forwarding mechanism which is not currently implemented in NDN testbed. They have adapted Poorman's solution for multipath. In this method, OSPFN permits operators to give a ranked list of all next hops and corresponding routes are inserted in FIB of

Fig. 4.5 Message
exchange process

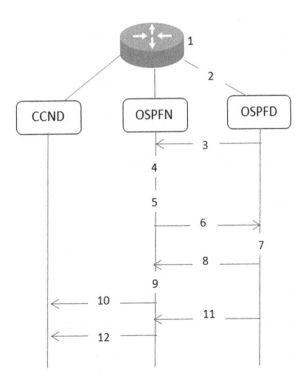

CCND. So, CCND can try other routes when the best route fails. Each route will be associated with preference and the one with the highest preference will be tried first. This multipath configuration will be specified for individual nodes, not for individual name prefixes, to ease the operator burden. This is just an initial sequence for CCND to explore; its forwarding strategy will select the best route that retrieves data in the fastest possible way. In configured multipath, OSPFN produces series of FIB entries for individual name prefixes. The most preferred next hops are those ranked with related route cost, followed by configured multipath next hops by decreasing order of preference. OSPFN will enter the FIB entries in reverse sequence as CCND will first try with the last-inserted FIB entries.

4.7   NLSR: A Secure Link–State Routing Protocol for NDN

The Named-Data Link–State Routing (NLSR) [10] is a distributed link–state protocol for NDN. NLSR has four key features. First is the use of a hierarchical naming scheme for routers, keys, and routing updates; second, the hierarchical trust model; third, is a hop-by-hop synchronization protocol; and finally, the forth is rank multiple routes for forwarding options. The router in the network comprises two modules: the NLSR and REPOSYNC module. The NLSR disseminates LSA periodically to prepare routing information in a distributed manner. A router sends a special message called Root Advice to neighbor routers that contain the combined hash value of the slice to the neighboring nodes. On receipt of the Root Advise, a router compares its slice with the incoming slice and replies if they are different. On completion of the synchronization process, the router sends Sync Notification to the local NLSR agent to fetch the data from the local repository and update the LSDB. NLSR protocol disseminates routing updates with the help of NDN's interest and data packets. To do so, a proper naming method for network routers, routing processes, and keys is highly essential. That is why NLSR has emphasized the naming mechanism for efficient routing. As network topology or name prefix can be changed at any moment, a routing update should be received promptly. As a link–state protocol, the NLSR incorporates identification of adjacency node and allows propagation of name prefix and connectivity information. Moreover, it generates multiple next hops for individual name prefix and uses proper signing and verification mechanism for all Link–State Advertisement (LSA) messages.

4.7.1   Naming Scheme

NLSR has incorporated an effective hierarchical naming scheme for efficient routing. This establishes a relationship among the components of the content locations. Each router has a name as per the network it resides in, the particular site it belongs to, and an assigned identifier of the router such as /(Network)/(Site)/ (Router). The router part contains two more information, the tag of the router and a label of a router. This name prefixes scheme helps to detect erroneous routing messages, to detect link failures by HELLO message exchange between neighboring NLSR routers, and to detect routing process failure itself.

4.7.2   Format of LSAs

Each router in NLSR gathers reachability and connectivity information with the help of LSA. An adjacency LSA is advertised by a router related to its links to a neighboring node. A name LSA keeps a list of name prefixes advertised by a router. Here, reachability leads to name prefixes that a node or its directly linked nodes can reach. Individual LSA are named as

/localhop/(network)/NLSR/LSA/(site)/(router)/(LSA-type)/(version)

The first field defines the scope; local hop restricts the LSA packet to be forwarded to immediate next hops only. The (*router*) field recognizes the router that produced the LSA. The (*LSA-type*) signifies either name or adjacency. The (*version*) field of a LSA is incremented by 1 when a router generates a new release of LSA. The LSA format for name and adjacency has been displayed in Fig. 5.4. Name LSA delivers all registered name prefixes locally to NLSR and those inserted by linked end nodes. An adjacency LSA describes all active connections of a router – each related with the name of neighboring router and the cost of link. This is generated at startup time of router and whenever any change in status of router's links occurs (Fig. 4.6).

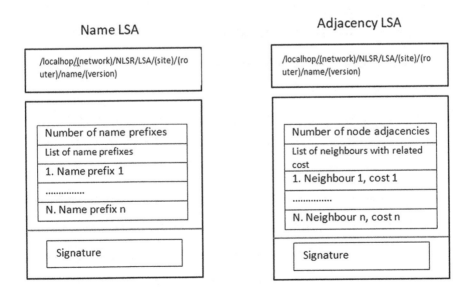

Fig. 4.6 The LSA format for name and adjacency

4.7.3   Dissemination of LSAs

A router propagates new version of its adjacency LSA to the network whenever it adds
or removes an adjacency with a neighbor. In addition, it advertises name prefixes from
dynamic registration as well as static configuration. A router will propagate a new
name LSA whenever any name prefix is added or removed. At each node, the link–
state database (LSDB) will maintain the latest releases of LSAs. On retrieval of any
new LSA, each router recomputes its paths and modifies the FIB. NLSR uses
ChronoSync protocol to synchronize router's LSDB updates. ChronoSync keeps all
the latest LSA names inside individual LSDB as a name set uses the hash of the name
set as an expression of the set. The difference in name sets, routers that run ChronoSync,
is identified by hash values of their respective LSA name sets. Instead of flooding a
new LSA to the whole network, if NLSR finds any new LSA name, ChronoSync
informs NLSR to obtain the relevant LSA. This separation allows the router to do
requests for LSAs whenever it has CPU cycles to avoid overloading of the router by
flooding of updates. The function of ChronoSync makes NLSR design simpler and
performance superior to IP-based link–state routing protocol.

Figure 4.7 depicts the LSA propagation inside an ICN topology. The ChronoSync
protocol in individual node keeps LSDB synchronized. It sends sync interest with

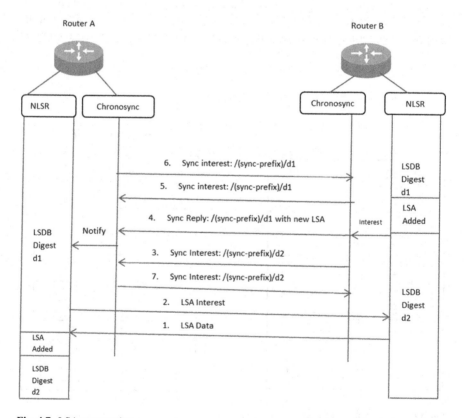

Fig. 4.7 LSA propagation

hash of all LSA names included in LSDB, to all nodes in network on regular time interval (steps 1 and 2). For example, if an LSA is added to LSDB of router B, then in B's LSA name set, the new LSA name is updated. B's ChronoSync also sends sync data packet carrying the new LSA name in response to sync interest from A (step 3). The ChronoSync at A retrieves the sync data, and informs NLSR regarding the new LSA name and modifies its LSA name set. Now, A and B recompute the hash for their name sets and send a new request for sync interest with new hash values (steps 4 and 5). As NLSR process at A has been informed about new LSA name, to retrieve this LSA, NLSR at A will send an LSA interest to B (step 6). B in turn responds with requested data of LSA (step 7). When NLSR at A retrieves the LSA data, it adds the LSA inside LSDB. At this point, LSDBs at both routers are synchronized. The link–state routing depends on LSAs for route calculation. In addition to this, there is no way for route calculation of LSA interests. Therefore, the LSA name prefix is configured with strategy called multicast forwarding, where LSA interest is sent to all the neighbors. If anyone has LSA copy in its CS or NLSR process, it sends it back. Otherwise, interest gets discarded because of local hop scope.

4.7.4   Adjacency Establishment

NLSR routers create an adjacency list before exchanging LSAs among them. This task is accomplished by sending hello interest initiated by NLSR module. The hello packet is transmitted at regular interval and the name of such packets are of the following form:

/(neighbour router)/NLSR/INFO/(this router)

The default interval time for sending the hello packet is of 60 seconds. It helps in identifying the status of a neighbor router. If the neighbor node responds with data that is signed by the neighbor's NLSR process key and can be verified based on the trust model, that neighbor is considered active or up. If the timer for hello packet expires, NLSR tries again by sending of interest in short intervals. After certain tries, neighbor is considered inactive or down. So, when an adjacency status changes, NLSR recreates its adjacency LSA and propagates it and routing table computation is scheduled. To discover repairing of failed links, NLSR keeps sending hello packets periodically and neighbor responds if a link is repaired. NLSR uses a face event notification through NDN Forwarding Daemon (NFD) to give quicker response to a failure of link. If the interface to the neighbor node gets failed, NFD notifies NLSR with a face ID related to that interface through face event notification. Then face ID is used by NLSR to compute the adjacent nodes that can be reachable via this interface, and marks that adjacency as inactive.

4.8   Recent Development in ICN Routing

Recent work on ICN routing is centered on NDN architecture. Since the NDN emphasizes purely content-based routing, all the proposed routing mechanisms have focused mainly on three attributes. To determine the closest copy of the content determined by the content name, searching and delivering the content via multiple paths (the multipath concept of IP routing), and stable and scalable content forwarding support. Few routing protocols as reported in different kinds of literature are described below.

A hyperbolic routing (HR) for NDN is reported in [11]. It uses a greedy mechanism for finding contents. Nodes in the network are modeled via coordinates in the hyperbolic space. Moreover, each node keeps track of the neighbors in terms of hyperbolic coordinates. The interest packets used for content discovery carry coordinates information to ease the content delivery. To forward objects or discovering, the algorithm computes the distance to their neighbor based on the coordinate information. The interest packets are forwarded to the closest next hop and hence propagates through the shortest path. The algorithm is capable of calculating the best path in real time with optimized use of the computational resources. It shows a stable topological behavior even in case of frequent link failure. Load balancing on the network is equally taken care of in the HR protocol. This work [12] presents an addressing scheme with optimally scalable routing. The proposed protocol needs the sizes of the FIB to be equal to the number of adjacencies that a node has. The routing overhead is minimal as no control messages for route establishment are used even for a drastic failure in the network. The addressing scheme is based on local information and only geographic coordinates of nodes in the topology are required. However, the protocol assumes that the network topology cannot be arbitrary and should follow an Internet-like topological design. Further, it is also suggested by the authors that the protocol can be most deployed in overlay networks. In case the topology grows, it must follow the predefined design specifications.

This research work [13] aims to improve the parameter related to user level performance like latency. Latency means the delay in time for content retrieval, once it has been requested. Here, they have proposed the concept of characteristic time to see improvement in latency parameter. The characteristic time defines the expected amount of time/likelihood in near future that content will be in the cache after it has been accessed recently. The proposed algorithm of characteristic time routing will use this characteristic time to send requests messages to the cache locations where there is a high probability that content will be found. The objective of the proposed algorithm is not to replace the existing routing strategies (Dijkstra's shortest path algorithm) but augment them and works in parallel with existing mechanisms of cache replacement and management, so it can be effectively implemented with the requirement of less effort in existing ICN testbed/Prototypes. the authors introduced a new data structure called "lookup table" to be maintained at each requestor, which will be first referred to fastest possible retrieval of content in order to decrease the delay. Furthermore, authors have suggested that in future, they will work for improving the working of the proposed algorithm with optimization of scope and granularity for the local search in lookup tables of neighbors while corresponding content name entry is not found in one's lookup table. In this situation, the requestor will look up in the neighbor's look up table to locate that entry with probability p. Authors also left the task of determining the p value depending on network overhead-delay tradeoff to future work.

The authors have proposed a controller-based routing scheme [14, 15] that runs on top of the name data networking paradigm so preserves all the features it. Due to named content and non-aggregated name prefixes, to store routes in NDN, exchange of huge amount of control packets is needed. The content replication at distinct location and mobility of content actually created an adverse effect on scalability. In order to overcome it, CROS will define particular names and functions for effective communication over NDN paradigm between routers and controller. The proposed approach will augment actions of router and decrease control and communication overhead between routers and controllers by coding the information related to routing on names of content. As proposed scheme supports requesting routes on demand by routers, it enables mobility of content and avoids creating replicas of routing data from controller to the routers. The proposed scheme also demands less memory for router, as it will store the routes for simultaneously consumed name prefixes. In addition, the proposed approach provides automation in router provisioning and installs a new path on every other router, in a path that has one route request toward the controller. The proposed protocol has been described using SDL, that is language for specification and description. And the same protocol behavior is validated using Petri Net in order to prove its correctness and feasibility. By simulation results, they have proved that efficiency of protocol is robust whenever interest rate of consumer increases along the extra throughput of more communication links and the efficiency is near-to-optimal value, provided routers operate with enough size of memory. They will extend their current work with other workloads and network topologies and evaluate the performance by exploring cooperation of routing as well as caching using the controller. The controller-based routing scheme for named data networking will address the issue of FIB explosion. When number of named data and related prefixes (non-aggregated) increases, the router need to store more routes and transfer more control packets that result in increased control message overhead and FIB explosion. The proposed scheme came up with specific controllers that are responsible for the following two tasks: a) get the network topology and compute routes; b) storage of locations for named data. The registration regarding location of named data will be done at controllers. Whenever an interest packet arrives for an unknown prefix, routers will send a request to controller to install a new route to that named data location. For the efficient distribution of storage of named data location, the proposed approach will use CRoS controllers to implement the distributed hash tables. So basically CRoS runs on top of the NDN paradigm and adapts all its features like controlling traffic, detect network problems, and diversity of paths. So, CRoS uses more than one controller to ensure scalability. Controllers will store the topology, compute routes, and storage of location regarding named data so that they will be able to install a path to any named content in the network. The decentralized functioning of the controller will redistribute the working load upon failure of any one or inclusion of anyone. The function of the controller permits reuses of router memory as it will store only the most accessed and recently accessed rules of forwarding. As the controller performs the division of a network into zones, the overhead related to flooding and route computation will be less. The authors will analyze, implement, as well as test the performance of the proposed approach in the future along with a comparison to existing related work. They will also investigate tradeoffs among patterns of route expiration and caching of routes.

The recent ICN routing protocols still emphasize the forwarding of interest packet to content producers without focusing on NDN's two prominent benefits that are multipath forwarding and in-network caching, which gives limitation to potential of NDN and advantages to its applications [16]. So, MUCA is a name-based new intradomain routing mechanism to give scalable and simple cooperation to network caching and multipath forwarding in NDN. It combines the advantages of link–state as well as distance–vector routing approaches. MUCA will first gather the topology information and calculate the shortest routes to content producers like any link–state protocol and it also gives more than one alternative route from neighbor routers like any distance–vector protocol. Also, it labels every routing update at the entrance of a network, so that all internal routers will choose that same border router for the identical name prefix. This will ultimately increase the cache hit ratio for the cached data. The authors have proved the efficiency of MUCA through simulation and conclude that it's effective to decrease the content retrieval latency and to increase the resiliency of network while decreasing routing overhead. MUCA will produce a new route, which is different than routine forwarding routes, to effectively utilize the in-network caching feature in NDN paradigm. MUCA will also propose a new method for synchronization of LSDB, in which the neighbor routers will be informed of incremental updates of routing. Finally, MUCA will fill the forwarding plane with a new list of ranked forwarding routes. The in-depth evaluation proves the superiority of MUCA over NLSR, the current routing mechanism implemented in testbed of NDN. MUCA's flexible design makes it suitable to accommodate any new routing solutions or ideas easily.

The authors have proposed a routing mechanism specifically for content-based network. A content-based network is a new communication paradigm where the packet will not be assigned an explicit receiver node address but the destination node for a packet will be discovered by matching the message content concerning selection predicates advertised by nodes. So routing process in this content-oriented network comprises predicate dissemination and the required information related to topology to preserve loop-free and minimum forwarding routes for packets. The proposed approach uses a mixture of traditional protocol for broadcast and a content-oriented routing protocol. The implementation requires a two-layer approach which, means a router will execute two distinct protocols, named broadcast routing procedure and content-based routing procedure. The result evaluation proves combined broadcast and content-based (CBCB) scalable and effective over existing routing solutions. In future, authors are planning to study parameters related to quality of service in content-oriented networks so that they can include this aspect while designing an improved version of forwarding and routing procedures.

A two-layer multipath routing protocol for NDN is reported in [17–19]. The proposal suggested to combine *Topology Maintaining* (TM) layer and the *Prefix Announcing* (PA) layer together to provide the routing support. TM layer maintains topology of the entire network and calculates the shortest path. The topology layer TM helps in finding the shortest path to the content by the service of upper layer. The Prefix, PA layer supports in actively publishing contents by disseminating announcements to all other nodes about the named content. It uses a single-source shortest-path tree constructed at the router for data transmission. Content routers keep their FIB table updated with the help of incoming content announcements. Likewise, in passive service mode, interest packets are flooded till it reaches to subsequent content provider or original producer. The proposed routing approach has certain flaws. First, publishing a content actively may increase the FIB entries. Secondly, passive service of the contents can increase the traffic in the network. These factors result in performance degradation and scalability issue. The authors of the work have proposed a popularity-based active publishing method to resolve such issues. In the enhanced version of the protocol, the upper PA layer publishes only the popular contents. The popularity is based on the access frequency of the content in the network. The unpopular contents are passively published by the protocol to reduce traffic load in the network.

4.9 Open Research in ICN Routing

An efficient routing mechanism plays a vital role in the successful deployment of ICN. Research community is continuously exploring this prominent area of research in the last decade. If routing fails, then there is a possibility of performance degradation. There are many research challenges concerning ICN routing. A few of such challenges are as follows:

(i) Node resource limitation: Due to the need for caching in ICN nodes there is a requirement for large memory. As the content in the network is finitely unlimited, hence providing accommodation for such a huge amount of content is challenging from a space point of view.

(ii) Scalability and overhead: Large number of contents and multiple cached locations in the network make it challenging to keep the routing scalable and also produce more overhead of content storing.

(iii) Find route efficiently: Due to caching of content, data may be acquired by many of the intermediate nodes in the network. In such a situation finding the closet node is difficult.

(iv) Trust and security: As the content may be delivered from any of the nodes other than the original producer, hence authenticity of content is challenging. Many of the routing mechanisms are trying to provide authenticity in coordination with naming system for ICN.

Table 4.1 Comparison of name-based routing

Reference	Timeline	Naming method used	Base ICN architecture	Multipath forwarding	In-network caching	Computational overhead	Fast reaction to failures	Routing update propagation problem?	Variation to base routing protocol (link state/dijkastra?)	Supports SDN	Mathematical model used	Simulation tool used
[7]	2012	URI based or CCNB	NDN	Support for configured multipath	Yes	Not addressed.	No explicit support.	Addressed	Link–state routing protocol-OSPF	No	No	NDN testbed
[9]	2013	Hierarchical	NDN	yes	Yes	Computational overhead is high.	Yes. resolved	Addressed	Link state routing protocol	No	No	Mini-NDN, an NDN network emulation tool based on Mininet.
[13]	2017	Not specified	CCN	Not mentioned, but can be	Yes	Less control overhead	No.	Not addressed	Link state routing protocol (Dijkastras algo)	No	Che's approximation/ Laoutaris' approximation to determine CT	Icarus simulator
[14]	2017	Hierarchical	NDN	Not specified	Yes	Less control and communication overhead	Yes	Not addressed	Not specified.	Yes	Specification and Description Language (SDL) and Petri Net	ndnSIM

Table 4.1 (continued)

Reference	Timeline	Naming method used	Base ICN architecture	Multipath forwarding	In-network caching	Computational overhead	Fast reaction to failures	Routing update propagation problem?	Variation to base routing protocol (link state/dijkastra?)	Supports SDN	Mathematical model used	Simulation tool used
[15]	2012	Hierarchical	NDN	Not specified	Yes	Less message and control overhead	Yes	Addressed	Dijkastra' algorithm	Yes	-	ndnSIM
[16]	2018	Hierarchical	NDN	Supported.	Yes	Lower	Yes.	Addressed	Link–state Distance–vector routing algorithm	No	-	ndnSIM
[18]	2004	Predefined packet format	Content-based network	Not specified.	Yes	Acceptable	Not specified.	Not addressed.	Dijkastra's algorithm	No	-	BRITE (Topology generator)
[11]	2016	Hierarchical	NDN	Supported.	Yes	Less compared to NLSR	No explicit support.	Not used.	Shortest path strategy with modifications	No	Probability concept	Mini-NDN
[12]	2017	Hyperbolic geometry-based addressing	NDN	Not specified.	Not focused.	Minimum – near to zero	Yes	Not addressed	shortest, geographic and geo-hyperbolic routing	No	Distance formulas for nodes in 3-D space.	NDN testbed
[19]	2016	Not specified.	CCN	Supported.	In network caching distance-based caching	Not addressed	Not specified	Not addressed.	named blind forwarding (BF) and provider-aware forwarding (PAF)	No	-	ndnSIM

Moreover, ICN has also been inherited from many applications like Internet of Things (IoT), Sensor Networks, or MANET. Mobility in such environment makes the routing task even more complex and brings many open research challenges like mobility and location management. Complete cooperation between routing and forwarding is also an important issue. Division of responsibilities and exchange of information between these two planes may decrease complexity of routing up to a great extent (Table 4.1).

4.10 Summary

The emergence of information-centric network paradigm has created quick attention from the various research groups, which treats content as first-class network citizen in place of the host. There has been less attention given to designing an efficient ICN-based routing solution that decreases the routing overhead and produce the routes that leads to cache hits by reducing content retrieval delay. This chapter begins with the explanation about significance of routing in ICN followed by exploration of existing ICN-based routing methods with its strengths and limitations. This chapter also gives a comparative analysis for all the existing routing mechanisms with respect to base architecture and several other performance parameters as mentioned in last section. The comparative analysis of state of the art conveys the need of an efficient ICN-based routing solution that can improve user- and network-level performance parameters by effectively utilizing benefits of ICN features.

References

1. Saxena, D., et al.: Named data networking: A survey. Computer Science Review. **19**, 15–55 (2016)
2. Yi, C., Abraham, J., Afanasyev, A., Wang, L., Zhang, B., Zhang, L.: On the role of routing in named data networking. Technical Report, NDN-0016 (2013)
3. Yi, C., Abraham, J., Afanasyev, A., Wang, L., Zhang, B., Zhang, L.: A case for stateful forwarding plane. In Proceedings of the ACM Conference on Computing Communication (SIGCOMM), Vol. 36(7), pp. 779–791 (2013)
4. Zhang, L., Estrin, D., Burke, J., Jacobson, V., Thornton, J.D., Smetters, D.K., Zhang, B., Tsudik, G., Claffy, K.C., Krioukov, D., Massey, D., Papadopoulos, C., Abdelzaher, T., Wang, L., Crowley, P., Yeh, E.: Named Data Networking (NDN) Project. [Online] (2010). Available: http://nameddata.net/project/annual-progress-summaries/.
5. Zhang, L., Estrin, D., Burke, J., Jacobson, V., Thornton, J.D., Smetters, D.K., Zhang, B., Tsudik, G., Claffy, K.C., Krioukov, D., Massey, D., Papadopoulos, C., Abdelzaher, T., Wang, L., Crowley, P., Yeh, E.: Named Data Networking (NDN) Project. [Online] (2011). Available: http://nameddata.net/project/annual-progress-summaries/
6. Zhang, L., Estrin, D., Burke, J., Jacobson, V., Thornton, J.D., Smetters, D.K., Zhang, B., Tsudik, G., Claffy, K.C., Krioukov, D., Massey, D., Papadopoulos, C., Abdelzaher, T., Wang, L., Crowley, P., Yeh, E.: Named Data Networking (NDN) Project. [Online] (2012). Available: http://nameddata.net/project/annual-progress-summaries/

7. Wang, L., Mahmudul Hoque, A.K.M.: OSPFN: an OSPF based routing protocol for named data networking. NDN Technical Report, NDN-0003, pp. 1–15 (2012). Online available at http://new.named-data.net/wp-content/uploads/TROSPFN.pdf
8. Van Jacobson, Smetters, D.K., Thornton, J.D., Plass, M.F., Briggs, N.F., Braynard, R.L.: Networking named content. In 5th international conference on Emerging networking experiments and technologies – CoNEXT '09, pp. 1 (2009)
9. Hoque, A.K.M.M., Amin, S.O., Alyyan, A., Zhang, B., Zhang, L., Wang, L.: NLSR: Named-data Link State. In Proceedings of the 3rd ACM SIGCOMM workshop on Information-centric networking - ICN '13, pp. 15–23 (2013)
10. Mahmudul Hoque, A.K.M., Am Syed Obaid Amin Adam Alyyan, S.O., Zhang, B., Zhang, L., Wang, L.: NLSR: Named-data Link State. In 3rd ACM SIGCOMM workshop on Information-centric networking, pp. 15–23 (2013)
11. Lehman, V., Gawande, A., Aldecoa, R., Krioukov, D., Wang, L.: An experimental investigation of hyperbolic routing with a smart forwarding plane in NDN (2016)
12. Voitalov, I., Aldecoa, R., Wang, L., Krioukov, D.: Geohyperbolic Routing and Addressing Schemes. ACM SIGCOMM Comput. Commun. Rev. **47**(3), 11–18 (2017)
13. Banerjee, B., Seetharam, A., Mukherjee, A., Naskar, M.K.: Characteristic time routing in information centric networks. Comput Netw. **113**, 148–158 (2017)
14. Torres, J.V., Alvarenga, I.D., Boutaba, R., Duarte, O.C.M.B.: An autonomous and efficient controller-based routing scheme for networking Named-Data mobility. Comput Commun. **103**, 94–103 (2017)
15. Torres, J., Ferraz, L., Duarte, O.: Controller-based routing scheme for Named Data Network. Technical report, Electrical Engineering Program, COPPE/UFRJ (2012)
16. Ghasemi, C., Yousefi, H., Shin, K.G., Zhang, B.: MUCA: New routing for named data networking. In IFIP Networking (2018)
17. Dai, H., Lu, J., Wang, Y., Liu, B.: A two-layer intra-domain routing scheme for named data networking. In: Proceeding of IEEE Global Communications Conference (GLOBECOM), pp. 2815–2820 (2012)
18. Carzaniga, A., Rutherford, M.J., Wolf, A.L.: A routing scheme for content-based networking. IEEE INFOCOM, 918–928 (2004)
19. Rehman, R.A., Byung-Seo, K.: Location-aware forwarding and caching in CCN-based mobile ad hoc networks. IEICE Trans. Inf. Syst. **99**(5), 1388–1391 (2016)

Chapter 5
Caching Mechanisms for Faster Content Retrieval

5.1 Introduction

As the ICN has greater importance for the content rather than the source (or producer more precisely) of the data, content caching is a vital concern. The more the contents are cached within the network, the more extensive is the possibility to have a less end-to-end delay and low network load. Thus, efficient cache management is a prominent issue in ICN. Cache management in ICN involves two primary concerns: the cache permission policies and cache replacement policies. The cache permission policies, also called caching strategies, determine whether a content needs to be cached or not and if content is to be cached, then where to cache the content. On the other hand, the cache replacement policy determines which content to be removed from the cache to accommodate new content if the cache storage is full. Caching is not a new topic of research and has been studied with regards to web caching [1–3] and CDN [4–6]. Although many of the issues of caching have been thoroughly analyzed in such fields, those solutions cannot be directly used in ICN. Since ICN emphasizes on the content rather than hosts, it prefers hop-by-hop transmission instead of the traditional channel-based transmission. The essential feature for designing the cache permission policy for ICN lies in the requirement of the line speed requirement for ICN routers. Furthermore, LRU is the default cache replacement policy in ICN, and it does not have the option to replace pages based on the access frequency. So, the complexity of the cache permission policy should be O (1) [7]. Moreover, the cache capacity cannot be enlarged arbitrarily and also it is determined by the bandwidth supported by the NIC. In addition, the ratio of the cache capacity and the content volume (in numbers) is in the order of magnitude of 10^{-5} [8].

© Springer Nature Switzerland AG 2021
N. Dutta et al., *Information Centric Networks (ICN)*, Practical Networking,
https://doi.org/10.1007/978-3-030-46736-4_5

5.1.1 General Overview of Network Caching

The prime objective behind caching data inside network is to increase the content availability to end users so that content requests can get satisfied by intermediate caches rather than visiting content source each time for needed content. By caching data in intermediate router caches, content retrieval delay can be reduced significantly. Caching has been integrated as an additional feature in ICN paradigm. This feature is capable enough to improve the Quality of Service (QoS) parameter experienced by end users, decreases the overall network congestion, avoids Denial of Service (DoS) attacks, and improves data availability.

Figure 5.1 shows the role of cache when any end user requests for content to network. Firstly, when the immediate content router receives the interest packet, it will store the incoming interface details through which interest packet arrives inside PIT table. Then it checks inside its own cache, located on right hand size in Fig. 5.1, to see whether it has cached the same requested content previously; if yes, the query will be answered by that cache itself to requester. The return path will be known to that router by referring to its PIT table. So likewise, all the intermediate content routers in ICN contain cache within it. Now if any subsequent request for same content comes to that router, it can directly send required data back to end user without visiting content source for the same. This action leads to significant reduction in data retrieval latency. This is a simple demonstration of how data availability can be increased by caching content inside network and how quality of service can be improved.

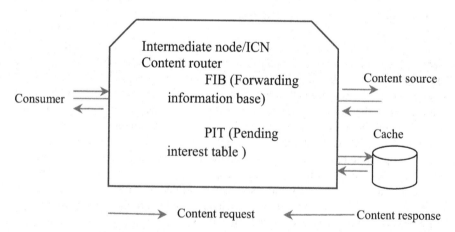

Fig. 5.1 Overview of network caching

5.2 Research Challenges and Issues in ICN Caching

Caching is one of the vital features to realize the benefits of ICN. Caching of data makes it possible to serve user request from anywhere in the network rather than contacting the source of data or producer. In the past decade, abundance of works is going on for providing an efficient solution for caching in ICN. However, there are several challenges [9, 10] that need further attention. This section focuses on few such challenges with a brief description of each.

5.2.1 Unique Naming of Content and Caches

Content identification by name is the prime need of ICN. For that, every piece of data is required to be named uniquely. Otherwise, same content may be cached in multiple names that leads to content deduplication. Hence, naming content uniquely is one of the vital concerns in the success of ICN caching. However, unique naming is highly challenging due to use of names other than numbers like IP addresses. There are many proposals found in the literature, but still the naming system has not matured. A suitable and efficient naming convention of content is a must for better ICN caching. Apart from naming content uniquely, we also require to name different caches present inside network uniquely. The naming of caches can be done by referring the name of the content router that contains it. Unique naming of caches is equally important to keep track of the caches that have content which has been requested by end user. This information can be useful to route upcoming similar content requests toward that cache. A content-based routing scheme in ICN may assume the identical name for content router as well as the cache that it contains. Therefore, an efficient naming scheme for content is much needed, and it is desirable to have unique naming for caches in ICN as well.

5.2.2 Popularity of Data

There is always a dilemma in deciding which content to cache and which to not. Research on traditional TCP/IP-based Internet architecture reveals many facts about the significance of usage pattern in data among current user interest and subsequently the probable benefits from networkwide caching. To mention a few, studies carried out in [9–11] have demonstrated the caching performance at the application level which is highly influenced by web information popularity. Hence, such popular data as revealed by the application must be cached to improve data delivery delay. However, every content available in the network cannot be cached and highly demanded content must be cache. It is suggested in many of the research work to cache the data that is identified as the most popular in the network. But the concept of data popularity itself is a vague term, and there is always a fluctuation in the popularity of data among users. Also, how to measure the popularity of data is another issue to be addressed. The heuristic information about a content being accessed is considered as one of the method to compute data popularity. In that case, the frequent change in the data popularity makes the computation complex and need frequent recomputation. That is why identifying the data to be cached is a challenging task in content caching for ICN.

5.2.3   Limited Storage Space in Caching Devices

It is very obvious that the Internet comprises of infinite number of contents that are supposed to be accessed by users. The ICN emphasizes on naming each of such contents uniquely. The caching on the other hand suggests to cache contents in the nodes of the network to improve content access efficiency. The most popular data can be cached inside in-network caches to reduce the content retrieval latency. Even if only the most popular contents are stored, such content is quite large in quantity and cannot be accommodated in the cache of network nodes. That is why managing the cache storage in the nodes of ICN network is another challenge. Moreover, various types of traffic competes for the same storing space, and hence it becomes significantly critical to perform cache management. Many recent research works have reported in this area of cache management, but there is still a need of further development in this area. The work reported in [12, 13] provides an insight in this issue along with certain potential solutions. They state that intelligent and flexible strategies can considerably improve the performance.

5.2.4   Storage Location

Another issue with caching in ICN is where to cache the content. There are many views in deciding the location to cache content. Some researchers suggest to cache the content in the end devices and some prefer to store in the intermediate routers in the network. The former is termed "edge caching," whereas the latter is called "in-network caching." Both the choices have their own merits and demerits. Furthermore, in-network caching is a simple technique to content placement during their transit from source to destination. That is why this technique is also called "on-path caching." The communication and computational complexity can be reduced using on-path caching but, conversely, the chances of hitting cached content items might also be reduced. In addition, in the network opens up the possibility easy cache management, routing, and forwarding. A suitably easy way to reach cache locations, indications of cache contention, and cache-ability of information can affect routing decisions in ICN [9, 10]. Keeping this in view, the concept of centrality betweenness is used to store a popular content. The betweenness centrality of a node in the network provides a measure of number of neighboring nodes connected to it. If a node have more numbers of neighbors connected to it then betweenness centrality value of that node will be high. Therefore, it will be more beneficial to store a content chunk at such router because it is reachable from a large number of other nodes. In short, proper selection of content caching location helps not only in easy access of data, it also supports in ICN content routing.

5.2.5   *Staleness or Freshness Detection of Content*

In most of the cases, the contents available on the Internet are dynamic in nature. This dynamism of data is only reflected in the source or the producer of the data. Hence, a cached content may not have its latest state as it is in the producer at the time of its access. Hence, it is at most necessary to know or preserve the staleness of data in ICN environment. Staleness detection of cached content is considered as another challenging task in ICN caching. It is because of the fact that the copies of named contents are largely distributed in in-network caches. The underlying ICN architecture must have the capability of providing staleness verification procedure for better synchronization of content publishers as well as in-network caching nodes. So far, there are two prominent approaches found in the literatures to resolve the said staleness issue: the direct and indirect approaches of freshness determination or staleness verification. The direct approach is suitable for some named contents, as this approach suggests that each copy of the cached content carries a timestamp regarding its time of caching. This timestamp directly specifies the freshness of the content in the cache. Based on the dynamism of the content in the recent past, the cached content may be used or erased from the cache as expired. On the other hand, the indirect approach does not provide any inherent information regarding its freshness. Rather, the content needs to be consulted about the staleness before transmitting it to a client. It happens in case of contents where it is difficult to set the expiry time of a content in advance. A webpage with dynamic interactive data is one such example. Where the main page is same, however, certain ads in the page or comments or review is dynamic [14].

5.2.6   *Repeated Caching of Same Content*

As we have talked about caching of popular content as determined by the inherent methodology to analyze data, the issue here is how frequently the popular content needs to be cached. Following only the case of storing data in cache, without considering the frequency of storing, leads to the method of copy everywhere caching mechanism. However, it produces too much redundant data in the network and in case of sudden change in the cached data, inconsistency becomes a big issue. Moreover, as discussed earlier, the storage is limited and the content to be cached is infinite. Hence, copying the content in each of the router on the path of its transmission leads to quick exhaustion of storage space and only few of the popular content would be cached. On the contrary, if data is cached sparsely, the search time of contents is more and hence it would not meet the objective of lower delay in ICN. Hence, it is highly essential to find an optimal solution to how often the same data should be cached. Research on this topic is still being carried out, with a wait for reliable and efficient solution.

5.2.7   Content Mobility

A study by Cisco Visual Networking Index [7] states that in 2021, the volume of data by mobile devices will raise by 18 times more than current. It has a big impact on edge caching approach used in ICN. Because if the edge caching is adopted, then end devices or user equipment are used for caching data. In such conditions, there is a high possibility that the selected user is mobile. Hence, based on the information hold by any intermediate device, it will send request to the end device that is caching the data. But by the time the client request reaches the cached location, the user might change its location and the request would not be served. It introduces considerable amount of overhead for searching the content. The said problem also occurs in in-network caching. Due to the inclusion of a large number of mobile devices in the network, the caching router may be mobile. In that case, the searching of contents will suffer from extra overhead due to absence of caching devices in its expected locations. The Mobile Ad Hoc Network (MANET) and Vehicular Ad Hoc Network (VANET) are examples of such environments, where all the devices including the router are mobile. It is important to consider the content mobility into account for implementing caching in ICN.

5.2.8   New Research Challenges for ICN Caching

There are plenty of existing ICN caching mechanisms available in literature till date. But there is a difference between ICN caching and conventional caching system. ICN is composed of a network of caches that are distributed at different geographical locations. There are two major objectives for ICN like distributed cache system:

1. To push data nearer to requestors so that user experienced parameter like Quality of Service can be improved.
2. To decrease the load on origin server as well as network bandwidth cost of data providers.

In order to achieve these objectives, cache providers also wish to utilize the cache resources efficiently. While designing the caching strategy and management schemes for cache systems, the objective should not only be to increase cache hit ratio for each cache, but also emphasize on the approach to utilize all the cache resources inside network efficiently [27]. It's unfortunate to say that classical caching schemes like least recently used (LRU) results in lower cache hit ratio at intermediate levels for hierarchical cache network as mentioned in [28, 29]. Some researchers disagree that data should be cached at edge nodes/servers only. But because of several constraints related to space, power, and bandwidth, cache servers are also placed at distinct hierarchical levels inside network. It is identified in [29] that the main problem for lower cache hit ratio and cache resource utilization on different intermediate caching locations is because of the problem called "thrashing" occurred by "filtered" request streams viewed by these nodes when caching mechanisms like LRU are implemented at each node independently. Therefore, the concept of "BIG" cache abstraction was introduced in [29] to manage and utilize cache resources of network fully and efficiently. In BIG cache abstraction, a complete line of distributed caches starting from an edge server till the origin data server are considered and handled collectively in the form of a virtual BIG cache in such a way that cache management schemes are applied to the whole virtual BIG cache in coordinated and coherent fashion. This allows any available cache replacement methods like FIFO, LRU, k-LRU to be used as a single consistent method to the whole virtual BIG cache. The major research challenge while designing BIG cache abstraction is to decide how to partition the cache resources logically at intermediate caches in order to allot proper cache resources to form single virtual BIG cache correspond to every edge server. To do so, distinct performance goals of data providers, users as well as cache network operators (CNOs) should be considered. Now to resolve this challenge, a network wide optimization model/framework must be developed that is built on the concept of BIG cache abstraction.

The problems of allotment of cache and content placement can be decoupled which leads to an optimization of decomposition framework in the context of management of cache network. In this, there will be an algorithm to solve both of these problems iteratively and separately. This framework allows to separately design efficient network-wide cache management in order to optimize the overall performance of cache network from point of view of data providers, CNOs, and users [27]. The primary challenge in designing any caching strategy for modern content distribution network is to become adaptive to the complete heterogeneity of the data that is accessed by end users. Nature of real-world data requests is a primary challenge in proposing a caching mechanism that automatically makes itself adaptive to the burstiness, dynamic, and heterogeneous. The authors in [30] have proposed adaptive TTL-based caching mechanisms with performance analysis that proves its superiority for these real-world content requests over state-of-the-art techniques. This adaptive TTL mechanism adapts a TTL value dynamically with the help of stochastic approximation. TTL value is used to find how long a content object may reside in cache. The major challenge in this caching strategy is to have a balance among two conflicting goals: maximize the cache hit ratio and minimize the cache size need. In this situation, it's favorable to permit distinct QoS guarantees for distinct categories of content objects. This means to have different cache size targets and cache hit ratios for different categories of content objects.

The researchers have also explored the way to integrate the name-based routing mechanism with caching technique in order to improve QoS parameters for network. The authors in [31] have proposed caching joint shortcut routing for ICN with similar kind of conception where shortcut routing (that leads to shortest route to the destination) with cooperative pre-caching method. Here, the focus of routing scheme is to reduce the overall route length and focus of caching method is to minimize data retrieval delay and traffic burden on content server. The authors in [32] have also proposed a dominating set-based collaborative caching with request routing in content-centric networking. Here, they have used dominating set-based caching method which is collaborative in nature. The proposed scheme creates a virtual backbone by applying connected dominating set theory to it. And then collaborative caching mechanism considers placement of content as well as request routing in ICN for this virtual backbone. This approach outperforms the existing caching mechanisms in ICN as per the simulation study mentioned in [32]. The major benefit of this kind of conception is that the caching strategy will place the most popular data inside core content routers and integrated routing scheme will forward the interest packet toward nearest core content router such that content retrieval latency can be minimized. The major challenge for this kind of collaborative scheme is to propose an efficient caching strategy first so that it can place/cache the most popular content at core routers intelligently and then embed it with an efficient request routing procedure to improve QoS parameters related to network.

5.3 Recent Trends in Caching

To ensure efficient network utilization and improve data availability, several ICN architectures make heavy usage of data caching. There are two major caching approaches: in-network caching and caching at the network edge. In-network caching is heart for caching mechanisms in ICN, and majority of researchers have utilized or proposed new mechanisms in this area of ICN caching. Despite this fact, there are new research directions for ICN caching to work upon or to borrow the existing caching methodologies from other well-established architectures to ICN (like network edge caching and SDN-based caching). In-network caching describes the caching of data within the transport network (e.g., on the forwarding path in network routers), or in conjunction with the Name Resolution Services. Caching at the network edge includes, for example, user nodes like in P2P networks and replicated servers. In this section a set of research proposals given so far is discussed briefly. Now, depending on the characteristics of caching mechanisms, we can decompose the available caching strategies into various categories [33]. Actually it is difficult task to find out a concrete criterion, we categorize them into below mentioned classes.

1. *Heterogeneous caching* vs. *homogeneous caching:* In case of homogeneous caching, the content routers will cache/store all the data messages travelled through that router, and each content router inside network has identical cache provision. While in case of the heterogeneous caching, each content router along the path from content source to requester may or may not store the data messages. In heterogeneous caching, each content router inside network carries distinct cache size. Let's say the content routers which bear heavy traffic load may have greater cache size so that it can cache more amounts of data messages.

2. *Noncooperative caching* vs. *cooperative caching:* The caching strategies can be divided into these two categories based on whether the content routers will cooperate with one another or not. In case of noncooperative caching, content routers will take caching decisions independently. It also does not advertise any information related to cached content to other routers inside network. But in case of cooperative caching, all the content routers inside network coordinate among each other to cache more data chunks. The content routers may create cache states, so that cached data becomes available to off-path content requests. Let's say, a forthcoming content request may be answered by a copy that has not been cached along the route from data source to the end user. Now based on whether the advertisement related to cache state exists or not, cooperative caching can be divided into two types: implicit cooperation and explicit cooperation. In case of implicit cooperation, the content routers do not require sending their cache states using an extra advertisement procedure. But additional operations are needed to help the routers for their forwarding or caching decisions. But in contrast to that, in explicit cooperation, the router should make aware others related to its own cache state by means of advertisement whenever router stores a piece of data within itself.

3. *Off-path caching* vs. *on-path caching*: Based on the location where data chunk is cached, caching strategies can be divided into off-path caching and on-path caching. In case of on-path caching, a data message is stored along its route to the end user. While in case of off-path caching, the data message may or may not be stored at routers along that route. The on-path caching is built-in for ICN, the off-path caching can also become possible if centralized topology manager comes into picture as it can take decisions related to where to store a data message and send a data message copy to the specific content router.

A discussion on caching approaches of ICN by exploring some of the promising caching solutions available in literature is as follows. This investigation is focused on the two fundamental caching mechanisms, namely leave copy everywhere (LCE) and Leave copy down (LCD) in detail and then rest of the existing schemes in brief.

5.3.1 In-Network Caching

A transparent, ubiquitous in-network caching can improve content availability and hence minimize the end-to-end delay and traffic load in an ICN. It also significantly alleviates the burden on network bandwidth that occurs due to rapid growth in traffic growth. Although caching has proven to be a useful method in current web and P2P networks for network load reduction, such methods are not suitable for ICN due to many reasons. In web or traditional Internet, there is a lack of unique identification objects and it leads to identifying same object with more than one name. There are ample of in-network caching protocols proposed so far for ICN. Different properties of content, like popularity, and characteristics of network topology, like betweenness centrality, etc., are taken into consideration to decide the content to be cached and location of caching. Moreover, with the advent of Software-Defined Networks (SDN), researchers are suggesting to adopt SDN to implement caching in ICN. This section describes few such protocols with a critical analysis of those schemes. At the end of the chapter, a detailed analysis of all these protocols are carried out and findings are presented in a tabulated form. But before going to recent development in caching, two most popular and highly used in-networking caching mechanisms, namely LCE and LCD, are discussed in this section as well.

5.3.1.1 Leave Copy Everywhere (LCE)

This is the simple mechanism for management of cache [4] in ICN since long time which suggests to cache the content in each node through which it travels. This approach supports the major goals of ICN proposal presented inside [10] to make content available easily and readily. For interest message propagation, they have considered hierarchical structure for content routers. The in-between nodes are arranged to have a cache hit. Fig. 5.2 demonstrates the LCE caching process. Here, $n_1, n_2, ..., n_m$

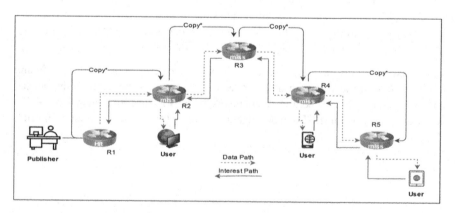

Fig. 5.2 Working of LCE

content routers are arranged hierarchically. The interest packet follows the same hierarchy during traversal. Let's say, node R5 (client) sent a request for content. This packet travels to R1, which is data publisher. Still, the desired content is found inside interior node n_i which has cached the requested content inside its cache. The copy of content is cached inside all intermediate routers on the path of data delivery in the sequence: $\{n_i-1, n_{(i+1)}-1, ..., n_{(1-m)}\}$. It ends with availability of transmitted content inside all the single-levelled network nodes in the route. It introduces data redundancy at the cost of increased data availability. As a result of this, we need an efficient cache replacement strategy as cache will be exhausted frequently. Various implementation of LCE is available along replacement strategy like First In First Out (FIFO), Most Frequently Used (MFU), Least Recently Used (LRU), Random, etc. Because of huge data redundancy, the LCE approach is not suitable with respect to ICN. This caching mechanism is homogeneous and noncooperative in nature.

5.3.1.2   Leave Copy Down (LCD)

This strategy of cache management presented in [4, 5] defines the way and procedure of data caching inside nodes. Its functioning is a similar like "drop at the first neighbor" method. In this method, once a request packet is posed, a route link is established with the form $\{n1, n2, ..., nm\}$ to the data publisher or content holder. As the data reports a cache hit on R1, it only saves the content copy inside its directly connected neighbor with the help of the traversing route to the R2 subscriber. For leave copy down, to get the high LCE states of data availability in almost every network nodes, it requires a data hit at every level of $\{n1, n2, ..., nm\}$. The functioning of LCD mechanism has been shown in Fig. 5.3. This caching mechanism is heterogeneous and cooperative in nature.

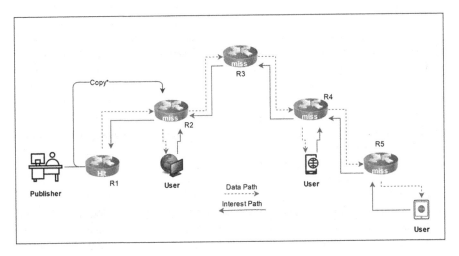

Fig. 5.3 Working of LCD

The authors in [2] presented a caching scheme that is based on local data popularity. It emphasizes on selecting suitable routers for content caching along with the set of contents to be cached. It uses local popularity of a content for content selection and betweenness-centrality concept to select a router for caching. The concept of centrality betweenness is taken into consideration to minimize the too frequent caching of same data as well as to make it reachable to large number of clients in the network. The centrality betweenness specifies the degree of neighboring nodes to a router. Higher the centrality betweenness, larger the possibility of being reachable in the network. If the content is cached in such node with high centrality betweenness, then a greater number of clients will be benefited by having an opportunity to access the cache. Caching scheme with similar concept is also found in [8] proposed by same author. The method proposed in [1] is a popularity-based content eviction scheme for CCN. As the research work mentioned in [2], it also takes into account the local popularity of a content and uses the betweenness centrality of routers. This work suggests to use a subset of routers in the network for caching. Hence, there is a possibility of overloading some routers, leaving others underutilized. Moreover, the concept of centrality is difficult to determine. Another popularity-based caching scheme is in [15]. The proposed scheme achieves high hit ratio by caching less content. A content is stored if the popularity reaches a threshold. It has the same disadvantages of finding an optimal threshold. The work of [8] is a cache replacement protocol based on least caching utility. Few of the nodes in the network are collaborative, and one of such node on the path evaluates the payoff of caching a content. They have designed cache utility function based on the content popularity and evict the content with least utilization. For collaboration among a set of nodes, collaborative messages are to be communicated which leads to extra messaging cost. However, the protocol reduces such overhead by piggybacking collaborative messages with data packets. The authors in [20] describe popularity-based caching policy as dubbed fine-grained approach. A popularity threshold is used to determine whether to cache a data chunk or not. Authors have also proposed to dynamically change the popularity threshold to reflect the changed popularity of content over time. The problem of threshold-based solutions is to find an optimal value. Another literature suggested in [5] presents a popularity-based caching strategy that suggests to cache the content when the popularity reaches a threshold value. However, the concern with these works is to find the suitable threshold. Too larger or small value of the threshold may lead to over or under estimation, because determining an optimum threshold is critical.

The authors in [4] have proposed a caching algorithm which suggests dividing contents in groups of clusters based on the type. Then popularity of each cluster is determined based on prediction. The protocol uses the autoregressive integrate moving average (ARIMA) model for popularity forecast. The caching strategy described here works as distributed without centralized controller. A major issue in this approach is the additional overhead incurred due cluster formation and grouping. The authors in [6] have suggested an algorithm to overcome the unnecessary duplication in caching content. It works on adaptive way to accommodate content on demand. This protocol suggests to cache content only in some selected nodes instead of all nodes on the delivery path. Also, in case of the cache full, the content is sent to the nearby node for caching. This process enhances the cache capacity by collaborating among nodes in storing content. Overhead is more due to implementation of prediction. This protocol has suggested using only a subset of routers in the network to cache. The algorithm overloads some routers leaving other routers underutilized. Moreover, the concept of centrality is a relative position and very difficult to determine. The authors in [7] have proposed an adaptive caching scheme for CCN. This protocol is based on fuzzy logic and suggests selection of routers for caching contents. This protocol uses betweenness centrality which specifies how many times a particular content (caching) router appears on the content delivery path among the all pairs of routers in a network. A caching router with high value of appearance is selected for storing the content. This scheme does not distribute the content equally over the network routers. Moreover, the overhead is large due to addition of extra fields in the interest as well as data packets and inclusion of training phase in the algorithm. The authors in [16] have done performance analysis of LFU and LRU algorithm in their work. The caches in both isolated and interconnected scenario are being extended with Che's approximation. The LRU and LFU are the oldest protocols for managing caches. It is extended to ICN by many researchers. In the work of [16], it is shown that the cost of LFU and LRU is high because they using flooding mechanism for caching and it consumes a lot of bandwidth. The authors in [17] in their paper have modeled the LRU cache policy by using Markov chain. They have calculated expected sojourn time as a function of content request rates and cache size. However, unless distribution parameters are chosen carefully, predicting sojourn time is difficult. It may lead to flooding if threshold is kept same as the lowest popularity. The authors in [18] have proposed a distributed content caching mechanism for ICN. The scheme is based on the probability approach that determines the available cache resources on the path. They have suggested to share same copy of content on the joint path. However, the protocol uses copy everywhere mechanism at the initial phase and hence produces huge redundancy. The authors in [19] proposed a method of efficient caching by utilizing SRAM and DRAM available in the router. A hierarchical structure of caches with DRAM lower and SRAM upper layer is designed provide two-layered, high-speed caching system. Due to involvement of hardware, significant speed performance is achieved. This caching scheme pre-fetches the content into SRAM and DRMA to reduce future access latency. However, such pre-fetching may increase the cache miss if content pre-fetching is not done efficiently. Moreover, router with other types of memory such as SDRAM, DDRA, or SSD RAM may not be efficient with this protocol due to the underlying speed limit of such memory types.

5.3.2 *Edge Caching*

The authors in [20] has described a cache replacement method using SDN support for applications in Fog-based sensor network. Decision of replacement is determined by age of the popularity, popularity of data chunk, and active period of the sensor node. Each of the data chunk is assigned a value based on these three parameters.

The higher-valued chunks sustain and lower-valued one is replaced. The proposed replacement strategy significantly improves the availability of precious sensing data in sensor network. Mobility in mobile node is not considered in their work. The authors in [21] in their work have proposed caching strategy with location prediction of content. It uses machine learning techniques to predict the interest of users on different content. This proposal is mainly focusing the content caching toward the edge of the network like consumer of data. They have proposed in their work an optimized replacement algorithm for improving the cache utilization. The authors in [22] in his article, provides a study on cache management policy along with routing mechanism for ICN. Many basic caching algorithms such as LRU, LCD, Randomly Copy One, etc., and routing protocols like Breadcrumbs, Hash-Routing, Characteristic Time Routing, etc. are studied in the paper. Some open research areas are also discussed in the article. However, he has not proposed any new protocol in this work.

The authors in [23] have proposed a caching mechanism for 5G heterogeneous network. Although it is not directly suggested to use in ICN. However it has many properties suitable to use in ICN. This protocol work has emphasized on saving energy of end devices by providing contents from nearby cached device. All nodes in the network cooperated with each other to provide its stored content to others. They have used generic algorithm to realize the concept of caching. As the use of intelligence technique in algorithms incur more overhead due to learning phase, the authors in [24] have presented an on-path and of-path caching for ICN. They suggest using a ICN Manager to create an end-to-end reverse path between the source and the user. For the purpose of caching, edge routers or on-path routers cooperates by optimally storing requested contents. A heuristic mechanism is used to offload and to optimally store the contents. However, previous knowledge of content retrieval is required to implement a heuristic approach.

5.3.3 SDN Based Caching

SDN is of the recent notable research trend in networking. It transmutes the traditional networks into service delivery architecture by decoupling control and data functionality of a routing process. This separation of functions isolate network infrastructure from user applications which in turn, facilitates the easy adoption of new network architectures in a more flexible way. With the increased adoption of SDN into traditional networks, researchers are trying to integrate it with different network implementations. It is noticed to be integrated into ICN as well. The synergy of ICN's efficient data dissemination and SDN's flexible network management is promising to design a fully controllable framework for efficient data communication. Primarily, researchers are coupling SDN with caching and routing in ICN to take advantage of data and control isolation in improving scalability and dynamism. In this subsection, few of the caching mechanisms in ICN caching is discussed briefly. Detailed discussion of SDN in general in the implementation of ICN is found in Chap. 8.

The research work mentioned in [25] has discussed an SDN-based caching mechanism for NDN. They primarily focus on the use of SDN in NDN architecture and have suggested a popularity-based caching mechanism. It calculates and attaches a prefix with the data and if the prefix exceeds a defined threshold, the SDN switch decides to save the prefix in table of content store to respond to future data requests. The authors in [3] having the same proposal is extended further for the same SDN-based caching mechanism. The suggested scheme considers Named Data Networking (NDN) architecture to implement the proposed SDN-based cache replacement scheme. It uses the popularity of data for caching in the network. The content is proposed to cache in SDN-enabled switches deployed in the topology. Before caching, the proposed protocol suggests to compute the popularity of the data chunk with an online popularity measurement technique, and the computed result is used to decide whether to replace a content in the cache storage or keep it for future use. This paper has studied the implementation of caching for a single zone in SDN network; however, behavior of the protocol is not studied for multiple zone data communication in SDN. So, there is a need of further analyzing the protocol. The authors in [26] have proposed a popularity-based lightweight caching scheme. It performs a prediction of popularity to decide caching of a specific content. It also integrates cache replacement with the placement scheme. This scheme assumes Software-Defined Networking (SDN) support by underlying network architecture and uses deep learning techniques. Due to the learning phase for popularity detection this approach incurs considerable computation overhead.

5.4   Performance Metrics Along with Objectives in Terms of Caching Mechanisms for ICN

In the existing research work related to ICN caching mechanisms, there is a huge variety of performance measures that have been adopted in order to analyze the performance of caching mechanisms. In this section, we will explore some of the fundamental and most common metrics that has been used for performance evaluation [34]. Along with discussion on individual performance measure, we will also state the objective for any caching mechanism in context of that performance measure.

5.4.1   Cache Hit Ratio and Server Load

In context of ICN, whenever an interest packet request gets satisfied by any cache inside network in place of the needed content's actual data producer, a cache hit happens. In contrast to that, when an interest packet requires to travel all the way to reach content producer, a server hit happens. Both the metrics of cache hit ratio and server load evaluates the percentage of interest packets that turns in cache hits or server hits, respectively. Therefore, these both metrics indicate that how well the popular data is disseminated across the entire network. So, the objective of any caching mechanism should be to increase the cache hit ratio as it will decrease the server load and therefore the burden on data producers will get decreased.

5.4.2   Content Retrieval Latency

The content retrieval latency evaluates the average time duration it takes for an interest packet to be satisfied; it includes possible retransmissions of interest packets as well. The content retrieval latency may be influenced by some of the caching-independent parameters, like network traffic and density, though it is also influenced by cache diversity. If the average availability of data chunks across the network is better, then the data retrieval latency will be less. If we make an assumption that network traffic and density will be approximately the same, we can utilize retrieval latency to investigate performance of caching mechanism in making data available quickly. So, the objective of any caching mechanism should be to decrease the data retrieval latency and to increase data availability across the network.

5.4.3   Retransmission Ratio for Interest Packets

The retransmission ratio of interest packets indicates the percentage of interest packets that are retransmitted because of interest packet time out. This performance parameter is related with content retrieval latency in that it is influenced by network traffic and density as well as an efficiency of caching mechanism. An ideal mechanism of caching can end up in cache diversity that is sufficient enough for interest packets to be satisfied always for initial transmission. The point to note here is that in ICN for the case of on-path caching, a retransmission does not compulsorily state that the content requires to be re-retrieved from the same place it was actually sent from. If the content was sent part of the way before it gets lost in-between, it may get cached in an intermediate router, therefore lowering the time needed to fetch it. In other way, the amount of extra load and latency caused due to given retransmission changes based on the cache's state on the path, and related retransmissions are supposed to be less expensive. So, the objective of any caching mechanism should be to decrease the total number of overall retransmissions for interest packets.

5.4.4   Total Cache Evictions

The total number of cache evictions shows how well the given caching mechanism is capable enough to adapt the cached data to the popularity and dissemination of data inside network. If a caching mechanism ends up in thrashing, this will become an evident for an increasing count of total cache evictions. In contrast, if a caching mechanism is capable enough to efficiently disseminate data inside network in such a way that caches can successfully satisfy as much interest packets as feasible, then the evictions of cache will be very rare. So, the objective of any caching mechanism should be to decrease the total number of overall cache evictions in network.

5.4.5   Diversity Metric

It is a performance measure that shows the diversity of data present inside caches throughout the network. It is represented by D; it is computed by following equation:

$$D = \frac{|C_{disj}|}{|S|}$$

Here $|S|$ shows the number of data producers and $|C_{disj}|$ shows the count of disjoint name prefixes available inside all caches. In this case, prefix is an identifier for data chunk producer. In other way, diversity metric calculates the percentage of data producers that are shown inside caches in the network at any particular moment. In ideal case, for a given application, if a cache diversity is a required objective than we want the value of DM to converge to 1. This means that every data producer has minimum one data object stored at any location inside network. The use of this performance metric can be essential in cases where diversity has high priority; it only computes the diversity in context of producers, not in context of actual data.

5.4.6   Cache Retention Ratio

Cache retention ratio is a content-level diversity measure. It actually complements the working of DM as it computes the ratio of different content objects that are present inside caches at any particular moment to all produced content objects; represented by C, this metric is defined by:

$$C = \frac{D_q}{D_p}$$

Here D_q gives count of unique data objects currently present inside minimum one cache and D_p gives total count of unique data objects produced throughout the lifetime of network. The point to be noted here is that as size of the cache is fixed, it is obvious that CRR will decrease as time progresses. This is because the new data objects are regularly being created and cache size remains the same. It means that data objects will gradually vanish from the network completely. This situation is expected; though, in context of comparing various caching mechanisms, it is interesting to investigate how fast CRR deteriorates. In other terms, the objective should be how efficient the proposed caching mechanism is at keeping data objects present inside network for as long as feasible.

So, these were all distinct performance metrics to evaluate any caching mechanism in ICN. We have discussed each performance metric with its significance and the corresponding objective in context of ICN. In next section, we will see a comparative analysis for existing caching mechanisms in ICN.

5.5   Analysis of Various Caching Methods

Caching is a hot research matter in ICN caching. ICN research is primarily concentrated in caching compared to other areas of research in this theme. In order to design new caching proposals, clear understanding of existing schemes is necessary. Keeping this in mind, Table 5.1 compares various caching policies that have been discussed so far. The table states different aspects of these methods with respect to model used, underlying network, simulation tools used, etc. The table will help readers to identify potential tools and mechanisms that may be used for new caching proposals. Comparison of existing caching mechanisms with respect to the year in which they have been proposed and several other parameters are mentioned below. In the table, tick mark denotes the support (Yes) for the mentioned parameter by given caching mechanism, while cross mark denotes that specified parameter is not supported (No) by given caching mechanism.

1. *In-network caching:* Caching contents inside content routers in order to increase data availability based on various in network caching schemes as discussed earlier.
2. *Edge caching:* Edge caching refers to the use of caching servers to store content closer to end users.
3. *Probability based:* In context of whether the proposed scheme has utilized the concept of probability to discover the cache that carries requested content.
4. *Popularity based:* In the context of whether the proposed caching scheme has utilized the content popularity parameter to take decision regarding caching of that content. The popularity of content signifies how frequently user requests for the same content again and again; in other words how demanding the content is among end users.
5. *Integrates cache replacement:* In context of whether the proposed caching mechanism incorporates any cache replacement strategy or not based on certain criteria related to cache.

Table 5.1 Comparative parameters

Reference	Timeline	In-network caching	Edge caching	Probability based	Popularity based	Integrates cache replacement	Flooding based	Prediction based	Supports SDN	Underlying Network support	Uses AI technique	Simulation tool used	α value for Zipf's Law	Reducing caching redundancy	Improving availability of cache	Centralized (C)/ Distributed (D)
[26]	2018	√	×	×	√	√	×	√	√	ICN	√(DL)	Mininet	0.1–1.0	√	×	D
[3]	2018	√	×	×	√	√	×	×	√	NDN	×	ndnSIM	0.5–1.5	√	√	D
[20]	2018	√	√	×	√	√	×	×	√	FOG	×	Ns3	No	×	√	D
[21]	2018	×	√	√	×	√	×	√	×	EDGE	√(ML)	Java	No	×	√	C
[6]	2018	√	×	×	√	×	×	√	×	ICN	×	ndnSIM	0.1–1.0	√	×	C
[7]	2018	√	×	√	×	×	×	×	×	CCN	√(ANF)	ndnSIM	0.4–2.0	√	√	C
[2]	2018	√	×	√	√	√	×	×	×	CCN	×	ndnSIM	0.4–2.0	√	×	C
[22]	2018	√	√	√	√	√	√	√	×	ICN	×	Icarus	0.6	×	√	-
[1]	2017	√	×	√	√	√	×	×	×	CCN	×	ndnSIM	0.4–2.0	√	×	C
[23]	2017	×	√	×	×	×	×	√	×	5G	√(GA)	None	None	√	√	C
[4]	2017	√	×	×	√	×	×	√	×	CCN	×	ndnSIM	None	√	×	D
[25]	2017	√	×	×	√	√	×	×	√	NDN	×	ndnSIM	None	√	√	D
[8]	2017	√	×	×	√	√	×	×	×	ICN	×	Icarus	0.6–1.4	×	√	C
[19]	2017	√	×	×	×	×	×	×	×	CCN	×	CCN caching	None	√	×	D
[24]	2016	√	√	×	×	×	×	√	×	ICN	×	Matlab	None	×	√	C
[5]	2016	√	×	×	√	×	√	×	×	ICN	×	Social CCNSim	0.7–1.0	√	×	C
[16]	2014	√	×	×	√	√	√	×	×	ICN	×	–	0.8	√	√	C
[17]	2014	√	×	×	×	√	√	×	×	ICN	×	Matlab	None	√	×	C
[26]	2014	√	×	√	√	×	×	×	×	CCN	×	Opnet	None	√	×	C
[15]	2013	√	×	×	√	×	×	×	×	CCN	×	ccnSim	1.5	×	√	C
[18]	2012	√	×	√	×	×	×	√	×	ICN	×	custom	0.8	√	×	D

6. *Flooding based:* In context of whether the proposed caching mechanism uses the flooding approach for content discovery and caching inside ICN, as the likelihood of finding the data can be increased by exploring a larger network area it means by flooding or collaboration.

7. *Prediction based:* In context of whether the proposed caching mechanism utilizes any prediction mechanism/algorithm while taking decision related to caching. These predictions will be useful in order to cache the most requested contents inside caches that are nearer to end users.

8. *Supports SDN:* The existing caching mechanisms can be differentiated based on whether it supports/works for the software-defined networking (SDN) framework or not.

9. *Underlying network support:* The existing caching mechanisms can be differentiated based on which underlying network paradigm it uses for base, like CCN, ICN, NDN, FOG, EDGE, 5G, etc.

10. *Uses AI technique:* The existing caching mechanisms can be differentiated based on which concept/strategy of artificial intelligence it has used in order to take decisions related to caching.

11. *Simulation tool used:* The existing caching mechanisms can be differentiated based on which simulation tool it has used for simulation of proposed approaches like ndnSIM, Ns-3, Icarus, etc.

12. *α value for Zipf's Law:* Zipfian is a distribution model that we have considered for incoming content requests from end users and α value denotes the distribution parameter related to it.

13. *Centralized (C)/Distributed(D):* The existing caching mechanisms can be differentiated based on whether it involves any centralized mechanism/ controller to take decisions related to caching or it works in distributed fashion.

14. *Reducing caching redundancy:* The existing caching mechanisms can be differentiated based on whether it resolves the problem of caching same content again at distinct caches or not.

15. *Improving availability of cache*: The existing caching mechanisms can be differentiated based on whether it maximizes the utilization of the cached content for their requested query without necessarily visiting the content server or not.

16. *Reducing caching redundancy*: The caching mechanisms can be differentiated based on whether they address the problem of caching the same content multiple times or not.

17. *Improving availability of cache*: The caching mechanism can be differentiated based on whether they have emphasized to increase cache availability so that cache content utilization increases and content retrieval delay decreases.

A comparative analysis of the existing caching mechanisms has been explored in Table 5.1. The difference between on-path and off-path caching strategies used in state-of-the-art literature, is also presented in Table 5.2. Various comparative parameters includes objectives, whether it is related to forwarding method or not,

Table 5.2 On-path caching vs. Off-path caching

Properties	On-path caching	Off-path caching
Objective	Leverage network congestion	Improve content availability
Relation with forwarding strategy	Strict	None
Background knowledge	Minimum/none	Needed sufficiently
Granularity of data naming	Data chunk or packet	Object
Validity of decision	Short duration (in seconds or hours)	Long duration (in days or months)
Location of caching	Intermediate routers	Servers (edge nodes)
Advertisement of data replica	None	Yes.
Content caching decision method	Heuristics	Optimization
Boundaries related to caching decision	None	Limited, example internet service provider or autonomous system.

granularity of data naming, validity of decision, location of caching, advertisement of data replica, caching decision boundaries, etc. (Table 5.2).

5.6 Open Research in ICN Caching

An important feature of ICN paradigm is to offer pervasive and transparent caching at interior nodes inside the network. This improves the utilization of network resources and content distribution speed. Because of rapid growth of network traffic, the task of caching content is a vital issue for ICN deployment. Following are the challenges that actually open future research directions for the research community.

- *Popularity of web application:* Different strategies related to placement and replacements of content have been thoroughly explored at application level mostly in context of web applications. The claims made in the paper [16] show that the benefits from exploiting ICN caching will not incur a desirable impact on throughput of the network. A recent analysis for the last decade has shown that the performance of caching at an application layer has been significantly modified by the popularity of web information.
- *Caching capacity and congestion type:* Whenever data chunks are cached within a network, different types of congestion try to access the same storage space. Due to this, the situation becomes difficult for networks to effectively manage existing cache storage. Recent research work on cache management has enlightened some flexible and intelligent methods that can effectively increase performance parameters related to caching in ICN.
- *Staleness identification*: To assure the freshness and consistency of cached data objects is also a challenging job for ICN caching. As replicas of the same data

chunk are disseminated across multiple caches of content routers, ICN must employ an efficient staleness assurance technique for NDO synchronization at publisher's site as well as at caches of interior nodes of ICN.

- *Data redundancy:* It is another vital issue that needs to be addressed in ICN. When the same content item is getting cached at each node inside the network, it creates duplicate copies of the same item. Majority of research related to ICN caching has addressed this issue but still the problem of caching analogous data chunks in on path content routers still exists.

- *Mobility:* Whenever a user inside a network, like Vehicular Ad Hoc Network (VANET) or Mobile Ad Hoc Network (MANET), moves from its current location, it interrupts all the related connections. This requires an up-to-date changes in routing table which actually introduces add-on difficulties for routing as well as name resolution. This future direction of work has gained less research interest and demands efficient solutions for the same.

5.7 Summary

In order to cope-up with the current Internet user's requirements, the information-centric networks and its associated architectures like NDN have gone through substantial modifications to make the content-centric approach adaptable to distinct network applications. This chapter explores the concept of caching in ICN along with its significance, fundamental types, challenges, and issues involved in it. This chapter also explores the existing caching mechanisms with its comparative analysis as well as strengths and limitations involved in each method. This chapter also highlights the need of an efficient caching method that can effectively utilize the ICN strengths and improve the network level as well as user-level performance parameters.

References

1. Kumari Nidhi Lal, Anoj Kumar, "A popularity based content eviction scheme via betweenness-centrality caching approach for content-centric networking (CCN), Wirel. Netw, 2017 pp. 1–12, 2017. https://doi.org/10.1007/s11276-017-1577-z
2. Kumari Nidhi Lal, Anoj Kumar, "A centrality-measures based caching scheme for content-centric networking (CCN)", Multimed. Tools Appl., vol. 77, no. 14, pp. 17625–17642, 2018
3. Kalghoum, A., Gammar, S.M., Saidane, L.A.: Towards a novel cache replacement strategy for named data networking based on software defined networking. Comput. Electr. Eng. **66**, 93–113 (2018)
4. Jiang, X., Zhang, T., Zeng, Z.: Content clustering and popularity prediction based caching strategy in content centric networking. In: IEEE 85th Vehicular Technology Conference (VTC Spring), 2017, Sydney, NSW, Australia (2017)

5. Hassan, S., Din, I.U., Habbal, A., Zakaria, N.H.: A popularity based caching strategy for the future internet. In: IEEE ITU Kaleidoscope: ICTs for a Sustainable World (ITU WT), Bangkok, Thailand, pp. 1–8 (2016)
6. Liu, W.-X., Li, J., Cai, J., Cai, J., Wang, Y., Liu, X.-C., Yu, S.-Z.: COD: caching on demand in information-centric networking. Telecommun. Syst., 1–17 (2018)
7. Lal, N., Kumar, S., Chaurasiya, V.K.: An adaptive neuro-fuzzy inference system-based caching scheme for content-centric networking. Soft. Comput. **23**(12), 1–11 (2018)
8. Huang, L., Yu, G., Zhang, X., Guo, Z.: On-path collaborative in-network caching for information-centric networks. In: IEEE Conference on Computer Communications Workshops (INFOCOM WKSHPS), Atlanta, GA, USA (2017)
9. Din, I.U., Hassan, S., Khan, M.K., Guizani, M., Ghazali, O., Habbal, A.: Caching in information-centric networking: strategies, challenges, and future research directions. IEEE Commun. Surv. Tutorials. **20**(2), 1443–1474 (2017)
10. Xylomenos, G., et al.: A survey of information-centric networking research. IEEE Commun. Surv. Tutorials. **16**(2), 1024–1049 (2014)
11. Ihm, S., Pai, V.S.: Towards understanding modern web traffic. In: Proceedings of the 2011 ACM SIGCOMM Conference on Internet Measurement Conference, pp. 295–312 (2011)
12. Psaras, I., Chai, W.K., Pavlou, G.: Probabilistic in-network caching for information-centric networks. In: Proceedings of the Second Edition of the ICN Workshop on Information-Centric Networking, pp. 55–60 (2012)
13. Chai, W.K., He, D., Psaras, I., Pavlou, G.: Cache 'less for more' in information-centric networks (extended version). Comput. Commun. **36**(7), 758–770 (2013)
14. Kutscher, D. et al.: ICN Research Challenges, 2015. [Online] Available: https://tools.ietf.org/html /draft-irtf-icnrg-challenges-03
15. Bernardini, C., Silverston, T., Festor, O.: MPC: popularity-based caching strategy for content centric networks. In: IEEE International Conference on Communications (ICC), pp. 3619–3623 (2013)
16. Martina, V., Garetto, M., Leonardi, E.: A unified approach to the performance analysis of caching systems. In: IEEE Conference on Computer Communications; Toronto, ON, Canada, pp. 2040–2048 (2014)
17. Liu, J., Wang, G., Huang, T., Chen, J., Liu, Y.: Modeling the sojourn time of items for in-network cache based on LRU policy. China Commun. **11**(10), 88–95 (2014) (Use this paper for Replacement algorithm)
18. Psaras, I., Chai, W.K., Pavlou, G.: Probabilistic in-network caching for information-centric networks. In: Proceedings 2nd ACM Information-Centric Networking Workshop (ICN), Helsinki, Finland, pp. 55–60 (2012)
19. Li, H., Zhou, H., Quan, W., Feng, B., Zhang, H., Yu, S.: CCNHCaching : high-speed caching for information-centric networking. In: IEEE Global Communications Conference (GLOBECOM), Singapore, pp. 1–6 (2017)
20. Al-Turjman, F.: Fog-based caching in software-defined information-centric networks. Comput. Electr. Eng. **69**, 54–67 (2018)
21. Tang, Y., Jianhua Mac, K.G., Yutong, S., Chi, T.: A smart caching mechanism for mobile multimedia in information centric networking with edge computing. Futur. Gener. Comput. Syst. **91**, 590 (2018)
22. Seetharam, A.: On caching and routing in information-centric networks. IEEE Commun. Mag. **56**(3), 204–209 (2018)
23. Ma, J., Wang, J., Fan, P.: A cooperation-based caching scheme for heterogeneous networks. Access IEEE. **5**, 15013–15020 (2017)
24. Rath, H.K., Panigrahi, B., Simha, A.: On cooperative on-path and off-path caching policy for information centric networks (ICN). In: IEEE 30th International Conference on Advanced Information Networking and Applications (AINA), Crans-Montana, Switzerland (2016)

25. Kalghoum, A., Gamm, S.M.: Towards new information centric networking strategy based on software defined networking. In: IEEE Wireless Communications and Networking Conference, San Francisco, CA, USA, pp. 1–6 (2017)
26. Mau, D.O., Chen, M., Taleb, T., Wang, X., Leung, V.C.M.: FGPC: fine-grained popularity-based caching design for content centric networking. In: 17th ACM International Conference on Modeling, Analysis and Simulation of Wireless and Mobile Systems, Montreal, QC, Canada, pp. 295–302 (2014)
27. Babaie, P., Ramadan, E., Zhang, Z.-L.: Cache network management using big cache abstraction. In: IEEE INFOCOM 2019-IEEE Conference on Computer Communications, pp. 226–234. IEEE (2019)
28. Fayazbakhsh, S.K., Lin, Y., Tootoonchian, A., Ghodsi, A., Koponen, T., Maggs, B., Ng, K., Sekar, V., Shenker, S.: Less pain, most of the gain: incrementally deployable ICN. In: SIGCOMM, 2013 (2013)
29. Ramadan, E., Narayanan, A., Zhang, Z.-L., Li, R., Zhang, G.: BIG cache abstraction for cache networks. In: ICDCS. IEEE (2017)
30. Basu, S., Sundarrajan, A., Ghaderi, J., Shakkottai, S., Sitaraman, R.: Adaptive TTL-based caching for content delivery. IEEE/ACM Trans. Netw. 26(3), 1063–1077 (June 2018)
31. Huang, B., Liu, A., Zhang, C., Xiong, N., Zeng, Z., Cai, Z.: Caching joint shortcut routing to improve quality of service for information-centric networking. Sensors. 18(6), 1750 (2018). https://doi.org/10.3390/s18061750
32. Xu, Y., Li, Y., Lin, T., Zhang, G., Wang, Z., Ci, S.: A dominating set-based collaborative caching with request routing in content centric networking. In: Proceedings of IEEE International Conference on Communications (ICC), pp. 3624–3628 (2013)
33. Zhang, M., Luo, H., Zhang, H.: A survey of caching mechanisms in information-centric networking. IEEE Commun. Surv. Tutorials. 17(3), 1473–1499 (2015)
34. Pfender, J., Valera, A., Seah, W.K.: Performance comparison of caching strategies for information-centric IoT. In: 5th ACM Conference on Information-Centric Networking (ICN '18), Boston, MA, USA. ACM (21–23 September 2018). [Online]. Available: http://conferences.sigcomm.org/acm-icn/2018/proceedings/icn18-final38.pdf

Chapter 6
Security in ICN

6.1 Introduction

Internet is getting populated by video traffic, and it was envisioned that 90% of such traffic shall be comprised of video by the year 2019 [26]. Due to proliferation of mobile devices and Internet of Things (IoT), the volume of video traffic is increasing rapidly. In IoT domain, every node can play the role of a provider; thus it leads to many-to-many communication paradigm. This increases size of routing tables and may lead to inefficient content delivery. Information-Centric Networking (ICN) paradigm is a result of research undertaken searching for efficient content delivery and scalable Internet [3]. ICN emphasizes on content-centric paradigm in which named content objects are decoupled from the hosts where they actually reside [24]. In the current host-centric paradigm, all requests for contents are made to the host where the contents reside, and such hosts are identified by IP addresses. Named content can be stored anywhere in the network. Each named object can be uniquely addressed and be requested to achieve by end users.

There are several ICN architectures available and to name a few are Named-Data Networking/Content-Centric Networking (NDN/CCN) [1], Data-Oriented Network Architecture (DONA) [4], Network of Information (NetInf) [5], and Publish–Subscribe Internet Routing Paradigm (PSIRP) [2]. These architectures differ in their details, but they share the same fundamental features such as contents are named uniquely, name-based routing, pervasive caching, and assurance of maintaining content integrity.

ICN improves delivery latency and at the same time gives rise to several security issues. Various security concerns in ICN includes denial-of-service attacks, different vulnerabilities in the context of ICN including cache pollution, content poisoning, naming attacks, etc. Different security attacks may increase frequency of retransmissions and finally the network throughput may be degraded.

© Springer Nature Switzerland AG 2021
N. Dutta et al., *Information Centric Networks (ICN)*, Practical Networking,
https://doi.org/10.1007/978-3-030-46736-4_6

6.2   Importance of Security in ICN

There are differences in security considerations for ICN and that of traditional host-centric networks. Content security is of highest importance in ICN. At the same time, channel security in ICN is of lesser importance in comparison to that in host-based networking paradigm. However, traditional security threats still exist in ICN. Various security threats such as snooping, Denial of Service (DoS), and impersonation attacks are very much prominent in ICN. Therefore, countermeasures to these attacks are necessary to understand and also relevant. The security issues in ICN have been addressed in this chapter. Different categories of security threats that have been considered here are as follows: Denial of Service (DoS), Content Poisoning, Cache Pollution, Secure Naming, Secure Routing, Secure Forwarding, and Application Security.

In ICN networking paradigm, there might be multiple copies of the same content spread across the network. This invites a difference between the security approaches that can be adopted for traditional host-centric networking and the same that can be applied in ICN. Unlike traditional host-centric networking, security cannot be bound to the end points or storage locations only, in case of ICN. Novel security techniques are required that can be applied to the content itself. Moreover, security measures are to be integrated inside the architecture itself and it cannot be left as an overlay on the architecture.

6.3   Key Security & Privacy Concerns in ICN Architectures

Information-Centric Networking (ICN) started with the TRIAD project initiated in Stanford University in the year 2000 [27]. There are several research projects related to ICN started thereafter. As mentioned above, Data-Oriented Network Architecture (DONA) [28], Network of Information (NetInf) [29], Publish–Subscribe Internet Technology (PURSUIT) [31], and Named Data Networking (NDN) [30] are few of those research projects to mention. All the above-mentioned ICN architectures have few concepts which are general and common in nature. Those concepts may be classified as follows: information object, naming, routing, caching, application programming interface, and security. Information object is nothing but the content itself. It is the major focus of ICN. It is important to note that each content may have different representations, and there can be different copies for each representation. *Naming* refers to the schemes by which contents are named. There are three different categories of naming schemes: hierarchical, self-certifying, and attribute–value pair based [6]. Unlike host-centric routing in which IP addresses are used, in ICN, routing is carried out using either name resolution or name-based approach. In name resolution, content name is resolved into a single or a set of IP addresses. And then request is routed to one of these IP addresses. In doing so, shortest path routing algorithm like Open Shortest Path First (OSPF) is used. On the other hand, in name-based routing approach, a request is routed based on the name of the content [32]. In network *caching* is a very important aspect in ICN. Caching follows the following principles in ICN: uniform, democratic, and pervasive. Uniform principle indicates that caching is applied to all contents delivered by any protocol. Democratic principle refers to the fact that caching policy is applied to contents published by any content provider. Finally, pervasive principle indicates that caching policy is available to all nodes in the network. Application Programming Interface (API) in ICN plays a very vital role in its operation. API is used to request and deliver contents in ICN. A source of content publishes its contents to make it available to others in the network. On the other hand, a source sends a subscription message in order to get the content in which the user is interested. These two operations "publish" and "subscribe" use the name of the content as the parameter to refer to the content. Security is a major concern in ICN. In principle, the network or any user of the network can use any available copy of the contents in ICN. Therefore, unlike host-centric networks, the security provisioning cannot be bound to the communication end points or to the storage locations. Thus, novel information-centric security concepts are required so that security can be guaranteed at the content level itself. It is not advisable to have a separate security layer as an overlay on the routing layer. Rather ICN architectures integrate security aspects within the architecture itself.

Key security concerns: ICN is unique in comparison to other technologies. There are at least five such attributes that make ICN unique. These attributes are mentioned below.

1. In ICN architecture, there is no host identifier. Thus, it becomes difficult to put limits on user requests.
2. The ICN architecture leads to the ubiquitous nature of the network, in the sense that any user can access any available copy at any location. This process adds difficulty to authorize user access.
3. Any user can publish/subscribe any content. This allows attackers to make faulty publications/subscriptions.
4. Various nodes in the network see the requests. This leads to the risk of losing privacy.
5. Security in ICN has to be integrated to the architecture itself. This is in contrast to a separate security layer as an overlay in many host-centric architectures.

ICN architectures face different attacks such as naming-related attacks, routing-related attacks, caching-related attacks, content provisioning attacks, and miscellaneous attacks. Miscellaneous attacks aim at degrading some ICN services. Here, attackers gain unauthorized access, and finally the attacks lead to insufficient or erroneous data distribution [6]. These are the various security concerns faced by ICN architectures.

Key privacy concerns: Privacy attacks in ICN are linked with the privacy matters associated with routers, cached contents, content names, content signature, as well as clients. All architectures of ICN face these privacy concerns. Privacy attacks may be classified as timing attack, communication monitoring attack, censorship and anonymity attack, protocol attack, and naming-signature privacy attack. Under timing and communication monitoring attack, attacker probe cached content of a router over time. Attacker identifies content popularity in the cache. Attacker also identifies requestor's content access behavior. Anonymous communication is important in ICN. Lack of anonymity can always reveal clients data and respective requested content. And at a later stage, this could be used to enable censorship. Thus, there is a risk of censorship if there is lack of anonymity. Another privacy attack, namely protocol attack, is special to two specific ICN architectures: CCN [26] and NDN [26]. Attacker can exploit design loopholes present in the protocol level of these two architectures. Pitfalls of the design features may be exploited by the attackers. In ICN, name of the content and signature on it bind the content to a particular producer (publisher). This fact raises privacy concerns of the producer (publisher). These are the various privacy concerns present in ICN.

In this article, the security concerns in ICN are focused on and thoroughly studied along with their countermeasures.

6.4   Attacks in ICN

ICN suffers from several security issues and they need to be addressed. There are few attacks in ICN which are new in nature. Such attacks did not occur before. Moreover, such attacks did not have any significant impact in other networking paradigms. In addition to that, there are some attacks that occur in other environments may also appear in ICN environments [8–13]. One such taxonomy as per [6] outlines following ICN attacks, considering both new and traditional attacks into four categories. These four categories are as follows: naming, routing, caching, and other miscellaneous related attacks. This classification is based on the attacker's main target.

It is important to mention that although each attack is included in only one category, such an attack may have impact in other categories as well. As an example, there are two attacks namely *flooding* and *unpopular request attack*, and these two attacks affect both ICN routing and caching. In a *flooding attack*, the main target of attacker is to overload and exhaust routing resources. As a consequence, this attack affects the caching system as well. Similarly, in *unpopular requests attack*, main target of the attacker is to violate cache relevance. Consequence of this attack affects the routing system as well. Various categories of attacks are briefly outlined in the following subsection.

A. *Naming-Related Attacks*: In ICN, the content requests are visible to the network. This fact increases the threat with respect to privacy. As the user requests regarding contents are visible, attackers control on information flow increases and they can block certain information flow with less effort. Attacker tries to prevent the distribution of a specific content under this type of attack. This is done by blocking delivery of the content [7, 14].

B. *Routing-Related Attacks*: In ICN, content delivery heavily depends on asynchronous publication and subscription. Effort is made to ensure data consistency among distributed data states. Two routing-related attacks namely *jamming* and *timing* aim to disturb state consistency. This may finally lead to unwanted traffic flows. Denial of service may also be a consequence. There are two more attacks namely infrastructure attack and flooding attack. These attacks try to exhaust the resources in the network hosts, like memory and processing power [15–19].

C. *Caching Related Attacks*: In ICN, caching is perhaps one of the important components because it plays an important role in delivering closest available copy of a desired content to a user. The overall performance of ICN infrastructure is heavily dependent on receiver-driven caching. Under this type of attack, attacker tries to pollute or corrupt the caching system [20–22].

D. *Content Poisoning Attacks*: In ICN, as already mentioned, entire focus is on content. Under this type of attack, an attacker tries to fill routers' caches with invalid content. In order to mount this attack, the control of one or more intermediate routers needs to be with the attacker, and then attacker can inject his or her own content into the network [26].

E. *Miscellaneous Attacks*: Objective of attackers under this category is to degrade some ICN services and also to gain unauthorized access. Such attacks finally lead to data distributions in the ICN which is insufficient and/or erroneous [23].

These five categories of attacks in ICN are detailed in the following sections.

6.4.1 Naming-Related Attacks

There are two types of attacks, namely *watch list* and *sniffing*, which fall under the category of naming-related attacks. In ICN, the network nodes can have access to the user requests. This fact is exploited by the attackers in order to launch these attacks. If an attacker can compromise an ICN node or a router, then it can access the user requests and also monitor the requester [14]. Since there is no host identifier in ICN, an attacker needs to compromise an ICN node or a router and then it can track requesters and also record who requested what [14].

Watch List Under this type of attack, an attacker maintains a predefined list of content names. This list is based on the interest of the attacker with respect to what is desired to be filtered or deleted. The attacker monitors the network links and performs a real-time filtering. The attacker now performs operations that may delete a request made by a user. Attacker may also record requester's information, considering and comparing with the predefined list. Moreover, the attacker may try to delete the matched content itself. Thus, an attacker launching watch list attack does the following: (i) the attacker may captures user requests; (ii) the attacker may filter and record who requested what; (iii) the attacker may filter and record return contents, which is basically the information about the publisher and the data; and (iv) the attacker may delete a request made by a user and also may delete contents. The entire operation set of the attacker is based on the predefined list maintained by the attacker.

Sniffing Under this type of attack, attacker does not maintain any predefined list as it was in the watch list attack. The attacker monitors the network. The attacker analyses the users' requests and also the content. The attacker searches for specified keywords in the requests as well as in the content, in order to make a decision regarding whether it is to be filtered or deleted. The attack scenario, regarding what an attacker can do launching this attack, is the same as the watch list attack.

Impact of the naming-related attack can be the following:

Censorship Attacker can censor the content as per his or her desire.

Privacy Attackers can monitor the requests made by the users. Attackers can also know about the requesters. Thus, privacy in the network is questioned.

Denial of Service An attacker can block a particular user's specific requests, for example. This block may be for some marked contents. Then this leads to a situation where requests made by a user remains unanswered. This is a kind of Denial of Service.

Content-naming scheme is an important aspect of ICN. In order to defend attacks like content poisoning, a verifiable binding between the content name and its provider should be provisioned.

Secure naming is another important aspect of ICN, and it can help in verifying provenance of a content. Similarly, secure routing and forwarding in ICN are also highly important aspects of ICN like any network architecture.

6.4.2 Routing-Related Attacks

There are several types of attacks related to routing in ICN. These attacks can be classified into two major categories namely Distributed Denial-of-Service (DDoS) attacks and spoofing attacks. Further, DDoS attacks can be of two types: resource exhaustion and timing attacks. There are four different types of resource exhaustion attacks, namely, infrastructure, source, mobile blockade, and flooding attacks. Similarly, spoofing attacks can be further divided into three different types of attacks: jamming, hijacking, and interception attacks. Figure 6.1 depicts these attacks related to routing in ICN.

Infrastructure Under this attack, the ICN may end up with Denial of Service. The scenario may be as mentioned below: an attacker sends a large number of requests

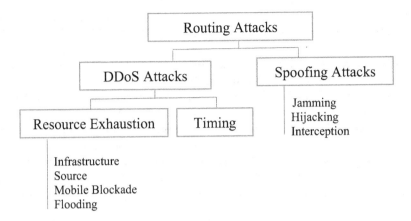

Fig. 6.1 Different routing-related attacks

for different contents which may be available or unavailable. These requests follow different routes in search of the source or availability of the content. In ICN, effort is made to find the closest copy of a desired content from the best available location. Due to large number of requests, the network may arrive at an overloaded condition. When the number of such requests is significantly high, it may lead to a Denial of Service. Moreover, users send retransmission requests after a limited and specified time duration. This may further amplify the problem.

Another scenario might be that the attacker may release large number of invalid requests. When the number of invalid request is significantly high, any legitimate request may take longer response time. As a consequence, if the response time exceeds a threshold level, that is, timeout period of the request, then the request may remain unanswered. This may lead to Denial of Service.

Source Under this attack, a single source is brought into an overloaded condition. Here, attacker sends a large number of requests to a specific content source considering a particular content. As a result, the source naturally takes longer service time and, therefore, there is increase in the response time with respect to the specified content delivery by the content source. This attack is launched to degrade the performance of a content source. This attack can lower the data return rate, and also it can affect requests made by all nodes available in the paths to the receivers. As a result, the overall network gets affected including the attacked sources.

Mobile Blockade Under this attack, a mobile attacker may overload a region of an ICN, by releasing large number of content requests. In order to launch this attack, the mobile user traverses neighboring networks on circular paths. The objective of the attacker is to overload the mobile access routers so that the state timeout period is exceeded, and finally blockade of the regionally available networks takes place [35]. In order to launch this attack, the attacker needs to be mobile, and the attacker needs to send a high number of requests to neighboring networks. Moreover, the attacker needs to traverse continuously and in a circular manner.

Flooding Under this attack, an attacker releases a significantly large number of requests. In an ICN architecture, the attacked node accepts a certain number of requests and then remaining requests are ignored. However, the attacker becomes successful in overloading the overall ICN infrastructure. Moreover, this harms all the proximate users. In ICN architecture, it is difficult to limit the request rate per end user because of the fact that host identifier is unavailable. As the attacker sends a large number of requests and that number exceeds the limits of the ICN nodes, it may so happen that the system neglects some of the legitimate requests directed to the attacked nodes.

Timing This attack works in increasing the request timeout for some ICN nodes. To launch this attack, an attacker releases a large number of requests in order to degrade the performance of some routers. This leads to a situation where request routing and data forwarding are compelled to exhibit longer delays. Here basic approach is to increase the request timeout for legitimate user's requests, by releasing a large number of requests through one or more routes.

Jamming Under this attack, a large number of malicious unnecessary content requests are sent by a node which is on a shared link. The attacker masquerades as a trusted subscriber. The attacker sends the malicious requests in order to disrupt the information flow in the network system. As a response to these requests, the ICN replies with the content which is sent to the destination having no receiver. Here the attacker sends requests to a shared node, which in turn forwards the same to neighboring nodes.

Hijacking One of the major differences between host-centric architectures and ICN is that contents in ICN can be cached and published/subscribed by any node. Under this attack, an attacker masquerades as a trusted publisher and then the attacker announces invalid routes for any content. These are in fact wrong information about content availability. Then if there are any content requests from users in the proximity of the attacker, then these are directed toward the invalid routes. Under this circumstance, naturally the requests will remain unanswered. This is another situation of Denial of Service. The effect of this attack may be significant if the attacker can hijack large number of invalid routes. However, the effect of this attack is lessened because of the fact that in ICN, routing mechanisms always route to multiple locations, and therefore, some legitimate routes may come up.

Interception Under this attack, the attacker who masquerades as a trusted publisher announces invalid routes, and also maintains a record of valid routes to the content. Then content requests are captured and diverted toward the proper location. The receiver gets the content as usual, however, the attacker gains knowledge of the requested content. This attack is similar to the usual "man in the middle" attack. Here, the attacker attracts users' requests by announcing invalid routes for some contents. From service-receiving perspective, the scenario seems to be normal but the attacker reveals the privacy of the users.

The attacks related to routing have impact on the following:

Denial of Service (DoS) DoS occurs due to many attacks under routing-related attacks category. For example, sending many requests for unavailable contents, sending many requests to a single source, mobile blockade, flooding, hijacking, and timing are some of the causes behind DoS. Intermediate timers delete requests with the expired timeouts, and this also leads to long delays and sometimes DoS.

Resource Exhaustion Some of the causes of resource exhaustion in ICN infrastructure are misuse of resources or generation of uncontrolled traffic in the network. Uncontrolled traffics are generated due to sending a large number of requests and the attacks like flooding.

Path Infiltration Unlike host-centric network, in ICN, copies of content are typically distributed to many untrusted locations. This makes it difficult to authenticate the origins for the contents. Under attacks like hijacking and interception, attackers claim themselves to be trusted and also announce invalid routes to attract user requests. These attacks are the major sources of path infiltration in ICN.

Privacy Under interception attack, users' privacy is violated. Attackers gain unauthorized access to user's requests and also to the contents.

6.4.3 Caching-Related Attacks

Various types of attack under the category of caching-related attacks are time analysis, bogus announcements, and cache pollution attacks. The cache pollution attack can be of two types, and these are random and unpopular request attacks. Figure 6.2 presents these attacks related to caching.

Time Analysis As per the philosophy of ICN, any node is eligible to cache any content. Under this attack, an adversary measures the time difference that occurs between request response times for cached and un-cached content [6]. Then this time difference can be analyzed to conclude if a proximate user has previously requested the same content as it has been requested by the adversary. Although it appears to be normal, because of this attack, adversary can gain information about such a proximate user. This fact violates the user's privacy.

Bogus Announcements It is noteworthy that caching system is a major part of the ICN architecture. Under this attack, an attacker can announce several times regarding

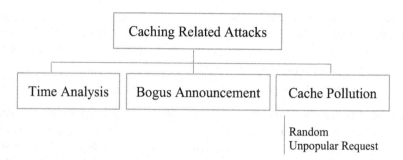

Fig. 6.2 Different caching-related Attacks

updates for content or cached copy. If the frequency at which such announcements are made exceeds the local content request routing convergence time, then it violates the caching and routing systems. As a result of this, it will be difficult for an ICN to match the legitimate requests with the appropriate content in the presence of these network updates which has been very frequent. Therefore, these overloaded announcements can lead to inappropriate and erroneous content retrieval.

Random Requests Under this attack, the objective of an attacker is to destroy the ICN in network caching system. Moreover, the attacker tries to change the content popularity. Thus, the attacker compels the ICN caches to store unpopular contents. The attacker sends random requests for these unpopular contents. Then the ICN caches these unpopular contents. A content that is not requested frequently is termed as an unpopular content. The attackers even may request false contents as a result of which the caches might get filled up with invalid contents. It is worth mentioning that a content is termed as fake if it is modified. Fake content does not come from the intended source; moreover, it is not the content requested by the user.

Unpopular Requests As it has been already mentioned, unpopular contents are not requested frequently. Under this attack, an attacker tries to destroy the ICN in-network caching systems and changes the content popularity. In order to do this, attacker requests unpopular contents. In order to become successful, attacker requires a prior knowledge about popularity of a content.

The attacks related to caching-related attacks have impact on the following:

Privacy User privacy is violated in time analysis attack, as the adversary may be able to know whether a user in the proximity has previously requested a particular content or not. The caching mechanism used in ICN is uniform, democratic, and pervasive [6]. This is one of reasons that invites greater privacy risks in comparison to the traditional network architectures.

Denial of Service (DoS) Bogus announcements can lead to DoS. Due to this, many frequent updates to contents may take place and it may lead to incomplete or erroneous data states. Users may not be able to retrieve desired contents, as mapping system will not be able to process the quick updates. And then this is a situation of Denial of Service.

Cache Pollution Random and unpopular requests, as mentioned above, can invite cache pollution [25]. Fake content, loading caches with unpopular content, etc. are the causes of cache pollution.

6.4.4   Content Poisoning Attack

Under this attack, the attacker tries to fill the caches of the routers with invalid content. In order to launch this attack, an attacker must gain control of one or more intermediate routers. By doing so, an attacker can inject own content into the network. Such injected contents shall have valid names with respect to an interest, but it shall be filled with fake payload or an invalid signature. Content poisoning attack can be launched in all types of ICN architectures. The ICN architectures using self-certifying names shall be less affected by this attack. This is so because in ICN with self-certifying names, the name of the packet and the digest of the packet's content match. This leads to the fact that the task of verifying the correctness of a content chunk, by comparing the hash of the chunk against the digest, becomes relatively easier. Finally, those packets whose hash does not match can be dropped. Due to this attack, an attacker can fill a network with poisoned content objects. Moreover, useful content may not find any place in the caches.

6.4.5   Miscellaneous Attacks

The attacks under this category are the following: packet mistreatment, breaching signer's key, and unauthorized access attacks.

Packet Mistreatment This is that type of active network attack which takes place during data transmission. The attacker having access to a link fraudulently or maliciously puts effort to block a communication, change the content, or reply to requested data many times. Under this attack scenario, the attacker gains accesses to ICN nodes or network links. As a result of this attack, the attacker may be successful in doing the following: modification of packets during transmission; replying to requester multiple times, leading to network overload and uncontrolled traffic; or generate content on behalf of a legitimate user.

Breaching Signer's Key Under this attack scenario, an attacker may launch any common attack to breach the signer's keys. In ICN, publishers sign contents in large volume which are going to be available for long time. Thus if attacker gains signer's key, there can be greater impact considering the security in the ICN.

Unauthorized Access   In ICN, multiple copies of the same content is distributed and made available in different network locations. In order to get unauthorized access, an attacker can use any one of the available copes for a particular content, and therefore it becomes easy for the attacker. As a consequence of this attack, an attacker gets access to a certain content sent to a specific user or group of users that the attacker is not permitted to access.

Miscellaneous attacks have impact on the following:

Congestion   The attacker may redirect the packets toward heavily loaded links. This may lead to congestion in the network. Moreover, packet mistreatment attacks can result in uncontrolled traffic in the network which eventually shall degrade the connection throughput.

Denial of Service   Under packet mistreatment attacks, the attacker can forward a significantly large number of packets toward a particular source. This obviously shall increase the network load, and therefore shall lead to a situation of Denial of Service.

Masquerading   If the attacker is successful in beaching signer's key, then the attacker may claim to be a trusted entity. And then the attacker can easily do the following: interception, analysis of the content, corrupt the communications, etc.

Unauthorized Access to Data   In ICN, routers are extremely important components. The routers have direct access to the content requests. The requests submitted by the users can be monitored by an attacker if the attacker gains access to a router by hacking it. Then a certain user can be tracked by the attacker by analyzing the user's request.

6.5   ICN Attributes Leading to Security Threats

There are few factors which lead to massive security threats in ICN. Various parameters that contribute toward vulnerabilities in ICN are as mentioned below: location-independent naming, ubiquitous publication/subscription, in-network caching, and state decorrelation. These attributes of ICN facilitate attackers to launch harder to detect attacks in the networks. Impact of attacks in the ICN may be severe due to these attributes. These attributes are detailed in the following section.

Location-Independent Naming   Contents in ICN are distributed into different locations, and these locations are not necessarily trusted. This permits content retrieval from any location which may be unknown or untrusted. Therefore, ICN needs a secure naming system in order to name contents irrespective of its location and representation.

Ubiquitous Publication/Subscription An ICN network can be accessed by any user from any location. Moreover, users can act as content suppliers or content consumers. Therefore, some users may send unwanted contents or even unwanted requests. Such activities may lead to performance degradation of the whole network.

In-network Caching In ICN, caching is very important and prominent. This characteristics of ICN architectures gives advantage to the users in one hand, and at the same time, it poses challenges. It is possible for any node of the network to cache any item that passes through it. Moreover, in order to satisfy the user requests, the contents can be delivered from the closest cache containing the requested contents, in the proximity of the user. There is no need to go to the hosting server in search of the contents.

State Decorrelation There are two asynchronous states that prevail in ICN, namely, request routing and content delivery. In ICN, there is a need of consistency between these two states. If the state consistency fails, it may lead to Denial of Service or unwanted traffic problems.

6.6 Security Mechanisms Adopted in ICN

There are few solutions adopted for ensuring security to various ICN architectures. However, this is an open area in which new developments are being awaited. In this section few such security solutions to specific security threats are highlighted.

Main focus in ICN is on content. The security solution in ICN should achieve privacy, censorship resistance, and plausible deniability. The security solutions should be of "easy to compute" nature for the users and computationally expensive for the attackers to break those.

6.6.1 Countermeasures for Naming Attacks

In [14], Arianfar et al. present a security solution to the naming attacks in content-oriented networks. The solution does not require shared keys to be maintained between the publishers and consumers. However, the solution does not provide ideal privacy to the users. The solution focuses on two attacks, namely name watch list attack and content analysis attack. This security approach demands substantial storage infrastructure. Moreover, the solution is based on several assumptions which may not be applicable to an instance of ICN.

In [33], Ion et al. proposes an encryption method based on attribute of the contents. Attribute-Based Encryption (ABE) has advantages over symmetric key encryption. In ABE, senders and receivers do not need to share any secret keys. With ABE, a receiver can decrypt data only when its decryption key satisfies the access control policies. These access control policies are embedded in the cipher text or the key itself. The basic ideas in this approach are mentioned below:

(a) Access control policies are attached to the data itself.
(b) Policies are enforced in a distributed manner.
(c) Policies are specified in terms of the content.

This approach has been implemented in Content-Centric Networking (CCNx) [34]. CCNx is based on NDN.

6.6.2 Countermeasures for Routing Attacks

In ICN, there is no host identifier. Therefore, it is difficult to put a limit on the number of requests generated by a host, in particular. An attacker can easily generate a large number of requests exceeding any specified limit. Finally, such an attack may lead to Denial of Service (DoS). ICN suffers from greater risk because such a network greatly depends on contents. Again contents can be created, deleted, or modified by any user of the ICN.

In [18], Alberto Compagno et al. proposes a countermeasure to Interest Flooding Attack (IFA). The solution consists of two phases: detection and reaction phases. Detection can be either local or distributed. In case of local detection, routers rely on local metrics only in order to detect an attack. These metrics are Pending Interest Table (PIT) usage, rate of unsatisfied interests, amount of bandwidth consumed to forward contents, etc. Distributed detection happens in a collaborative manner. Nearby routers collaborate and detect whether an attack has happened or not. Routers also collaborate to mitigate such attacks. This security solution is applicable to NDN. In [19], Afanasyev et al. propose few alternate solutions to the IFA. The solutions suggest to limit the request rate with a constant function. The solutions exploit the fundamental principle of NDN architecture, which may be stated as follows: "Flow balance between Interests and Data Packets." Flow balance essentially means "one Interest can be satisfied by at most one Data Packet."

In [16], Gasti et al. provide details about various Distributed Denial-of-Service (DDoS) attacks in NDN. There are few routing attacks also mentioned which lead to DDoS. Corresponding solutions are also suggested. In [17], Fotiou et al. propose ranking algorithms applicable for ICN contents. This algorithm is based on publisher and subscriber ranking. The solution can fight with spam.

6.6.3 Countermeasures for Caching Attacks

In ICN, caching happens to all contents and in all nodes. Most frequently requested contents should be stored in caches. In [21], Mohaisen et al. propose a privacy protection mechanism aiming at time analysis attack. Different caching policies are not considered in this solution. It has also been assumed that adversary is situated at the proximity of the attacked user. However, it may not be the case always. In [20], Xie et al. propose a mechanism (named CacheShield) to protect against cache pollution attack (random and unpopular requests). There are two major components in this solution: (a) a probabilistic shielding function, and (b) a vector of content names and their corresponding request frequencies. Scalability aspect of this solution needs to be evaluated.

6.6.4   Countermeasures for Content Poisoning Attacks

Countermeasures for content poisoning attack may be categorized as follows: (a) collaborative signature verification, and (b) consumer-dependent approach. In collaborative signature verification, routers collaborate to verify the content signature. In consumer-dependent approach, mechanisms rely on using additional fields either in request and data packets, or in client's feedback.

(a) **Collaborative signature verification**: In [16], Gasti et al. propose a countermeasure against cache poisoning attack. Routers can validate received content chunks by using "self-certifying interest/data packet," (SCID). This method is computationally less expensive than traditional RSA-based signature verification. However, the mechanism incurs extra overhead due to communication and storage of data, thereby limiting its scalability. Due such overheads, content retrieval latency is increased. In [35], Kim et al. uses check before storing (CBS) [36] concept. This approach probabilistically verifies the content, and only validated items are stored in the caches. It is measured that approximately 10% of the contents are requested again before their expiry in the caches. Therefore, the cache is divided into "serving content" and "bypassing content." Serving contents are those which shall be requested while they are in caches, and bypassing contents are those which shall be dropped from their caches before subsequent interests. Segmented Least Recently Used (LRU) policy is used for cache replacement. This approach can prevent from caching poisonous contents. However, the method is computationally expensive and, therefore, increase latency. Scalability is another issue associated with this method. At a higher scale, there can be significant negative impact on system efficiency.

(b) **Consumer-dependent approach**: In [22], Ghali et al. address the content poisoning attack. The solution provided here is intended for NDN architecture. A ranking algorithm based on consumer feedback is used to distinguish between valid and malicious content by a router. The major drawbacks in this approach are: the entire method is heavily dependent on client's feedback. Therefore, noncooperative client or malicious feedback may undermine the effectiveness of the method.

6.6.5 Countermeasures for Miscellaneous Attacks

The existing solutions for miscellaneous attacks cannot directly be applied to ICN because of its unique characteristics. However, there are few solutions with respect to miscellaneous attacks as classified but aiming at smart grid data collection. And such solutions can be investigated for ICN architectures. In [37, 38], Kim et al. propose a scalable and secure transport protocol (SSTP) along with end-to-end message protection aiming at smart grid data collection. These symmetric key-based and lightweight implementations are possible at client as well as server side. In [39], Kim et al. propose a secure platform (SeDAX) for smart grid communication. The proposed solution is scalable that can handle high volume of data as expected in ICN. In [40], Vieira et al. present a secure protocol for the ICN specially designed for smart grids. However, key management in this protocol needs more investigations. These security solutions perhaps may be applied to ICN with some modifications. It is noteworthy that ICN does not depend on conventional client–server architectures. Therefore, authentication server may not be feasible to implement from scalability perspective. Moreover, in ICN, it may be difficult to depend on shared key as any user can publish or subscribe any content.

6.7 Open Research in ICN Security

The ICN architecture design inherently extends its support for various privacy and security features like identity privacy and provenance. These are still not adequately present inside current internet paradigm. Still, provided its origin, the ICN architecture contains several open privacy and security concerns. Here, we present some open questions and research directions in the field of ICN security like security threats, privacy risks, and access control enforcement mechanisms.

The proposed mechanisms so far in the context of secure routing and naming for ICN paradigm are good initial attempts to resolve the possible malicious attacks in ICN. An efficient content naming method along with a verifiable binding among the content identifier and its producer is necessary to mitigate network attacks like content poisoning attack in ICN. Though, in all existing mechanisms [7, 41–43] this binding incurs high cost related to signature verification (full verification of binding needs signature verification for every content chunk). This would actually restrict intermediate content routers from verifying signature for each incoming packets in order to maintain line speed. Still, a computationally efficient and scalable approach needs to be proposed. An efficient and secure naming method is still an open research direction. It should encapsulate metadata, like data hash, the publisher's identity, and signature for purpose of advanced security. This is recently a prominent research area with various proposals submitted to the ICN Research community named Internet Research Task Force [44].

In addition, secure forwarding and routing do not perform acceptably for the scenario that involves consumer and/or producer mobility. Though this has not been addressed in the literature, exploiting Bloom filter-driven routing (z-filter) in publisher/subscriber networks may lead to a potential ICN routing attack. If intermediate router does not authenticate Bloom filter, a malicious content router can alter the bits inside filters in order to overload the entire network or to disrupt the process of content delivery. Designing an efficient strategy to help content routers to validate the authenticity and integrity of the z-filters requires major research attention.

6.8 Conclusion

Due to the significant increase in digital contents and Internet-based applications, there is a need to re-engineer Internet architecture or at least internal technologies in the years to come. Information-centric networks (ICN) are an effort toward that change. Such networks are fundamentally different from traditional host-based networks, in the sense that in ICN entire focus is on contents. Contents are replicated and made available in different locations so that users can easily achieve those avoiding long traverse and also traffic. However, security challenges are more in ICN due to its nature. In ICN, the tasks like content-based routing, in-network caching, and content naming are significantly challenging which are not necessary in traditional host-centric networks. Making ICN secure is a challenging task. In this chapter, various security attacks are discussed. Various denial of service attacks are highlighted. The attributes of ICN that lead to security threats are also outlined. Along with time, as the ICN paradigm achieves maturity, it is inevitable that new security challenges are also to be faced. It has been observed that while designing security solutions for ICN, in order to make ICN secure, the security concepts or techniques should be integrated in the internal architecture level and not as an external layer of security. Ensuring privacy and access control in ICN are also highly important aspects which have been beyond the scope of this chapter.

References

1. Jacobson, V., Smetters, D., Thornton, J., Plass, M., Briggs, N., Braynard, R.: Networking named content. In: Proceedings of the ACM International Conference on Emerging Networking Experiments and Technologies, pp. 1–12. ACM (2009)
2. Tarkoma, S., Ain, M., Visala, K.: The publish/subscribe internet routing paradigm (psirp): designing the future internet architecture. In: Future Internet Assembly, pp. 102–111. IOS Press BV, Amsterdam (2009)
3. Ghodsi, A., Shenker, S., Koponen, T., Singla, A., Raghavan, B., Wilcox, J.: Information-centric networking: seeing the forest for the trees. In: Proceedings of the ACM Workshop on Hot Topics in Networks, p. 1. ACM (2011)

4. Koponen, T., Chawla, M., Chun, B., Ermolinskiy, A., Kim, K.H., Shenker, S., Stoica, I.: A data-oriented (and beyond) network architecture. ACM SIGCOMM Comput. Commun. Rev. **37**(4), 181–192 (2007)

5. Ahlgren, B., D'Ambrosio, M., Marchisio, M., Marsh, I., Dannewitz, C., Ohlman, B., Pentikousis, K., Strandberg, O., Rembarz, R., Vercellone, V.: Design considerations for a network of information. In: Proceedings of the ACM CoNEXT Conference, pp. 66.1–66.6 (2008)

6. AbdAllah, E.G., Hassanein, H.S., Zulkernine, M.: A survey of security attacks in information-centric networking. IEEE Commun. Surv. Tutorials. **17**(3), 1441–1454 (2015)

7. Dannewitz, C., Golic, J., Ohlman, B., Ahlgren, B.: Secure naming for a network of information. In: Proceeding of the IEEE INFOCOM, pp. 1–6 (Mar. 2010)

8. Djenouri, D., Khelladi, L., Badache, A.: A survey of security issues in mobile *ad hoc* and sensor networks. IEEE Commun. Surv. Tutorials. **7**(4), 2–28 (2005)

9. Polla, M.L., Martinelli, F., Sgandurra, D.: A survey on security for mobile devices. IEEE Commun. Surv. Tutorials. **15**(1), 446–471 (2012)

10. Mpitziopoulos, A., Gavalas, D., Konstantopoulos, C., Pantziou, G.: A survey on jamming attacks and countermeasures in WSNs. IEEE Commun. Surv. Tutorials. **11**(4), 42–56 (2009)

11. Igure, V.M., Williams, R.D.: Taxonomies of attacks and vulnerabilities in computer systems. IEEE Commun. Surv. Tutorials. **10**(1), 6–19 (2008)

12. Xiao, Z., Xiao, Y.: Security and privacy in cloud computing. IEEE Commun. Surv. Tutorials. **15**(2), 843–859 (2013)

13. Hansman, S., Ray, H.: A taxonomy of network and computer attacks. Comput. Secur. **24**(1), 31–43 (Feb. 2005)

14. Arianfar, S., Koponen, T., Raghavan, B., Shenker, S.: On preserving privacy in content-oriented networks. In: Proceedings of the ACM SIGCOMM Workshop ICN, pp. 19–24 (Aug. 2011)

15. Vahlenkamp, M., Whlisch, M., Schmidt, T.C.: Backscatter from the data plane – threats to stability and security in information-centric networking. Comput. Netw. **57**(16), 3192–3206 (Nov. 2013)

16. Gasti, P., Tsudik, G., Uzun, E., Zhang, L.: DoS & DDoS in named data networking. In: Proceedings of the 22nd International Conference on Computer Communication and Networks, pp. 1–7 (2013)

17. Fotiou, N., Marias, G.F., Polyzos, G.C.: Fighting spam in publish/subscribe networks using information ranking. In: Proceedings of the 6th EURO-NF Conference on NGI, Paris, France, pp. 1–6 (June 2010)

18. Compagno, A., Conti, M., Gasti, P., Tsudik, G.: Poseidon: mitigating interest flooding DDoS attacks in named data networking. In: Proceedings of the IEEE 38th Conference on Local Computer Networks, pp. 630–638 (Oct. 2013)

19. Afanasyev, A., Mahadevan, P., Moiseenko, I., Uzun, E., Zhang, L.: Interest flooding attack and countermeasures in named data networking. In: Proceedings of the IFIP Networking Conference, pp. 1–9 (May 2013)

20. Xie, M., Widjaja, I., Wang, H.: Enhancing cache robustness for content centric networking. In: Proceedings of IEEE INFOCOM, pp. 2426–2434 (2012)

21. Mohaisen, A., Zhang, X., Schuchard, M., Xie, H., Kim, Y.: Protecting access privacy of cached contents in information centric networks. In: Proceedings of SIGCOMM, Hong Kong, China, pp. 1001–1003 (May 2013)

22. Ghali, C., Tsudik, G., Uzun, E.: Needle in a haystack: mitigating content poisoning in named-data networking. In: Proceedings of SENT, San Diego, CA, USA, pp. 1–10 (2014)

23. Fotiou, N., Giannis, G.F., Polyzos, G.C.: Access control enforcement delegation for information-centric networking architectures. In: Proceedings of the Second Edition of the ICN Workshop on Information-Centric Networking, pp. 85–90 (Aug. 2012)

24. Tyson, G., Sastry, N., Rimac, I., Cuevas, R., Mauthe, A.: A survey of mobility in information-centric networks: challenges and research directions. In: Proceedings of the NoM, New York, NY, USA, pp. 1–6 (June 2012)

25. Gao, Y., Deng, L., Kuzmanovic, A., Chen, Y.: Internet cache pollution attacks and countermeasures. In: Proceedings of the 14th IEEE ICNP, pp. 54–64 (Nov. 2006)
26. Tourani, R., Misra, S., Mick, T., Panwar, G.: Security, privacy, and access control in information-centric networking: a survey. IEEE Commun. Surv. Tutorials. **20**(1), 566–600 (2018)
27. Cheriton, D., Gritter M.: TRIAD: a scalable deployable NAT-Based Internet architecture. Stanford University, Stanford, CA, USA. [Online] Available: http://ceng.anadolu.edu.tr/cakinlar/BIL555/icerik/2000-Triad.pdf
28. Koponen, T., et al.: A data-oriented (and beyond) network architecture. SIGCOMM Comput. Commun. Rev. **37**(4), 181–192 (Oct. 2007)
29. The network of information: architecture and applications. SAIL, Brussels, Belgium, FP7-ICT-2009-5-257448-SAIL/D-3.1 (Jul. 2011). Accessed on 16 June 2020
30. Jacobson, V., et al.: Networking named content. In: Proceedings of the CoNEXT, pp. 1–12 (Dec. 2009)
31. Lagutin, D., Visala, K., Tarkoma, S.: Publish/subscribe for internet: PSIRP perspective. In: Towards the Future Internet, vol. 4, pp. 75–84. IOS Press, Amsterdam (2010)
32. Ahlgren, B., Dannewitz, C., Imbrenda, C., Kutscher, D., Ohlman, B.: A survey of information-centric networking. IEEE Commun. Mag. **49**(7), 26–36 (July 2012)
33. Ion, M., Zhang, J., Schuchard, M., Schooler, E.M.: Toward content centric privacy in ICN: attribute-based encryption and routing. In: Proceedings of ASIA CCS, Hangzhou, China, pp. 513–514 (Aug. 2013)
34. Palo Alto Research Center (PARC): Content Centric Networking (CCNx). https://www.ccnx.org/
35. Kim, D., Nam, S., Bi, J., Yeom, I.: Efficient content verification in named data networking. In: Proceedings of International Conference on Information-Centric Networking, pp. 109–116. ACM (2015)
36. Bianchi, G., Detti, A., Caponi, A., Melazzi, N.: Check before storing: what is the performance price of content integrity verification in LRU caching? ACM SIGCOMM Comput. Commun. Rev. **43**(3), 59–67 (2013)
37. Kim, Y., Kolesnikov, V., Kim, H., Thottan, M.: SSTP: a scalable and secure transport protocol for smart grid data collection. In: 2011 IEEE International Conference on Smart Grid Communications (SmartGridComm), pp. 161–166 (2011)
38. Kim, Y., Kolesnikov, V., Thottan, M.: Resilient end-to-end message protection for large-scale cyber-physical system communications. In: 2012 IEEE Third International Conference on Smart Grid Communications (SmartGridComm), pp. 193–198 (2012)
39. Kim, Y., Lee, J., Atkinson, G., Kim, H., Thottan, M.: SeDAX: a scalable, resilient, secure platform for smart grid communications. IEEE J. Sel. Areas Commun. **30**(6), 1119–1136 (July 2012)
40. Vieira, B., Poll, E.: A security protocol for information-centric networking in smart grids. In: Proceedings of the SEGS, Berlin, Germany, pp. 1–10 (Nov. 2013)
41. Wong, W., Nikander, P.: Secure naming in information-centric networks. In Proceedings of the Re-Architecting the Internet Workshop, p. 12. ACM (2010).
42. Zhang, X., Chang, K., Xiong, H., Wen, Y., Shi, G., Wang, G: Towards name-based trust and security for content-centric network. In Proceedings of IEEE International Conference on Network Protocols (ICNP), pages 1–6 (2011)
43. Hamdane, B., Serhrouchni, A., Fadlallah, A., S.G. El Fatmi, S.G.: Nameddata security scheme for named data networking. In Proceedings of IEEE International Conference on the Network of the Future (NoF), pages 1–6, (2012)
44. Information-Centric Networking Research Group. https://irtf.org/icnrg.

Chapter 7
Optimization in ICN

7.1 Introduction

Internet is getting populated by huge numbers of contents every day. The current service model of Internet may not be appropriate in the days to come to handle huge data and satisfy users with quality of service. Therefore, novel networking paradigm such as Information-Centric Networking (ICN) has been one of the most widely studied networking topics in recent times. The basic paradigm shift is that ICN decouples sender from the receiver. The ICN has caching capabilities in the network. Contents are named and distributed across the network; therefore, these are made available in many locations. When a user requests a particular content, it is retrieved from the location closer to the user. This approach reduces network traffic, delivery delays, and increases Quality of Experience (QoE) for the end user [18]. Internet Service Providers (ISP), Content Providers (CP), and Content Distribution Networks (CDN) are going to get new roles and business opportunities due to such networking paradigms [1]. These are the reasons why ICN architecture may replace present host-based networks in the future.

There are several protocol-level issues raised by such networking paradigm. Content naming, content routing, and in-network caching are some of the important issues. Optimal routing and optimizing location of various contents are important aspects to be considered while upgraded network performance is desired.

In this chapter, the issue of optimization in ICN has been addressed. The necessities of optimization, existing approaches for having optimized ICN, and future research directions in the line of optimization in ICN are highlighted.

© Springer Nature Switzerland AG 2021
N. Dutta et al., *Information Centric Networks (ICN)*, Practical Networking,
https://doi.org/10.1007/978-3-030-46736-4_7

7.2 The Need of Optimization in ICN

Considering a host-based system, optimization at various levels, such as routing, medium access, etc., are necessary in order to have better throughput in the network system. Similarly, in ICN too optimization is required to achieve upgraded service quality. For this, various parameters, such as delay involved in content delivery and routing cost, etc., are desired to be reduced. Finally, optimization is required to place contents in various locations in ICN and also in routing those contents toward the end users, so that congestions can be avoided and network traffic is also minimized. As the Internet is getting populated by multimedia traffic at very huge rate, meeting the end user–level expectations is a challenge. Handling video data is highly resource consuming. In order to have high performance of video delivery systems, even if ICN is used, it is necessary to have optimal routing and optimal caching strategy in place.

Energy has always been a constrained resource. Therefore, optimal use of this resource is highly important. It is desired to have minimum energy consumed while distributing contents in the network. In ICN, the sources of energy consumption involve mainly two processes. These are (i) content request made by the user, and (ii) the content transmitted to the requestor user. The sum total of the respective energy consumed by these two processes is the complete energy consumption between the two nodes. The energy consumption between two nodes is closely related to the distance between the nodes. The network energy consumption model establishes the relationship between the distance (physical separation) between the nodes and the energy consumed. This lays foundation toward having optimal solution considering energy as the parameter to be optimized, while placing the contents in in-network caches. Therefore, optimization is required with respect to the energy consumption generated by the distribution of contents as well as reception of contents between nodes [2].

7.3 Approaches for Optimization in ICN

There are several instances of ICN namely, PSIRP [3], DONA [4], NDN [5], CCN [6], etc. Content caching is an optimization problem. This problem has been investigated in different contexts on the Internet. For example, Li et al. studied optimal placement of web proxies for networks. They used a tree topology [7]. In [8], authors have proved that the optimization problem of CDNs is nondeterministic polynomial-time (NP) complete. In [9], Qui et al. addressed the problem of placing web servers in optimal locations for real network topologies. However, the solutions to such problems in the web or CDNs are not suitable to apply in ICN, for example in CCN. The unique characteristics of CCN are as follows: (a) *content needs to be located by its name and not by its location*, and (b) *every ICN node can cache and also serve a content if it is requested.*

There are few works available in the line of optimizing cache placement in ICN. In [10], Psaras et al. studied caching policies based on Markov Chains. In [12], Rosensweig et al. proposes an approximate model for performance analysis of CCNs. In this model, contents are supposed to be cached at each node along the path while requested contents are being made available to the end users. In [13], Rossi and Rossini address cache allocation problem for individual CCN routers. In their solution, centrality metrics such as betweenness, closeness, and degree centralities are used. In [14], Araldo et al. studies the problem of reducing cost in caching, in which a cost-aware cache decision policy is proposed.

There are few works available that consider both caching and routing in ICN together. In [15], the cache allocation problem for CCN has been studied. Here, the problem is formulated as a 0–1 maximization problem, following the structure of knapsack problem. This is based on a simplified model in which bandwidth is unconstrained and routing is fixed. In [16], adaptive mechanisms are studied to manage content replication as well as routing in continuous manner. In [17], optimization of content caching and routing are addressed considering a hierarchical tree network. In [11], the joint problem of routing and caching in ICN (CCN as the instance) has been addressed. Here, any node can cache and share content. The problem of CCN caching is considered as a two-stage stochastic programming problem. Finally, it is transformed to a deterministic multi-scenario linear program. The major characteristic of this formulation approach is that it is deterministic evaluation of cost and constraints in a nondeterministic scenario that involves the uncertainty of content request.

In [11], the problem of joint routing and caching has been stated as mentioned below:

It has been understood that it is necessary to optimize the replication of content objects in ICN, as it is possible to cache as well as share contents among the nodes. Optimal replication of content objects shall minimize routing costs, and it shall be feasible to function under bandwidth constraints. Now if the problem of joint routing and caching is to be stated, it can be done as follows:

If a particular cache location is considered, it is important to solve the problem of optimal routing under bandwidth constraints of the network system. The problem of optimal routing converges to select which node should serve a particular content object, requested by another node, in such a way that all content requests made by different users are satisfied and the total transmission expenditure is minimized. The expenditure of a potential solution for optimal cache location is considered to be the cost of optimal routing. Hence, the problem of finding optimal cache location reduces to the job of finding a solution to locate content objects that optimize the transmission expenditures under the optimal routing strategy. Optimal routing strategy is constrained by limited bandwidth and limited storage capacity.

Formal statement of the problem is as follows. This portion is borrowed from [11]. Various notations used are as follows:

h is the intermediate node,
r is the root node,

I is the set of STB nodes,

$I_1 = I \cup \{h\}$,

and $I_2 = I \cup \{h, r\}$

J is the set of content objects. Without loss of generality, It is assumed that content object j_1 is more popular than content object j_2 if $j_1 < j_2$.

n is the total number of STB nodes,

m is the total number of content objects,

S_I and S_h are the number of content objects that can be stored at a STB and the intermediate node respectively,

C_o is the capacity of the uplink (I, h) for each $i \in I$,

[C_o is a small integer, specifying the maximum number of content objects stored at i which can be uploaded to other customers during a given time period (e.g., the peak hour time period in a typical day). A downlink capacity is not specified because it is assumed that in all realistic scenarios for demands, the downlink capacity is sufficient to download the content objects required by any customer i (either from r, from h, or from another customer i')].

$y = (y_{ij})$ $(I \in I_1, j \in J)$ is a candidate solution of cache location, where node i stores content j if $y_{ij} = 1$, or not if $y_{ij} = 0$.

$d = (d_{ij})$ $(i \in I, j \in J)$ is content demand in a given scenario of demands where d_{ij} is a 0–1 random variable. If node i requests content j, $d_{ij} = 1$, otherwise $d_{ij} = 0$. The various random variables d_{ij} are assumed to be independent.

w_1 and w_0 are associated transmission costs when a content object is transmitted by a link between r and h, and a link between h and $i \in I$, respectively. It is assumed that $w_1 > w_0$ due to the fact that the connection between the intermediate server and the ISP network is long distance and it uses an expensive technology in data transmission.

For content object j required by customer $i \in I$, the routing cost of satisfying that content request by node $i' \in I_2$, denoted by $w_{ii'j}$, is.

0 if content object j is available at i (i.e. $i' = i$, $y_{ij} = 1$),

w_0 if it is downloaded from h,

$2w_0$ if it is downloaded from another customer $i' \neq i$,

$w_0 + w_1$ if it is downloaded from r.

Here, for $i \in I$, $i' \in I_2$, and $j \in J$, the routing decision variable $x_{ii'j} \in \{0,1\}$ denotes whether or not content j required by node i is delivered from node i'. The cost of routing of a possible routing solution x, for a scenario of content demand d, and content caching y is expressed as follows:

$$\varphi(x,y,d) = \Sigma\Sigma w_{ii'j} \, x_{ii'j} \, d_{ij}$$

Now the problems are stated formally, as mentioned below.

Problem 1 (***Routing Problem***): Given cache location (y_{ij}), where $i \in I_1$ and $j \in J$, and a scenario of content demand (d_{ij}), where $i \in I$ and $j \in J$, find a routing solution ($x_{ii'j}$), where $i \in I$, $i' \in I_2$ and $j \in J$, satisfying all the requirements of the customers in order to minimize routing cost $\varphi(x, y, d)$ subject to constraints on uplink bandwidth.

Suppose that we know the probability distribution of the possible scenarios of demands. Specifically, for each customer i, content j is required with a given probability pij is greater than or equal to 0. For a given probability distribution of demands, we denote $\psi(y)$ the expectation of routing cost with respect to a feasible caching location y. Then, the optimal caching problem is defined as follows.

Problem 2 (***Caching Problem***): Given the probability distribution of content demand, find $y = (y_{ij})$, where $i \in I_1$ and $j \in J$, in order to minimize $\psi(y)$ subject to constraints on storage capacity.

The above formal statement of the problem is borrowed from [11]. This has been incorporated to increase the readability of the chapter. Now reader can develop solution to the above-mentioned optimization problem. However, a solution to the same problem is available in [11] itself.

7.4   Machine Learning for Optimization in ICN

Machine learning (ML) is the area of computational science with the emphasis on structure analysis in data in order to help in future prediction [19]. It supports engineers and researchers to learn the behavior of a system and to make decisions based on the predicted results. Nowadays ML is applied in almost all types of applications, including healthcare, energy, production, market forecasting, and many more. In computer networking also, ML is used in various technologies like IoT, WSN, ad hoc network and so on. Needless to say, that the ML is suitable for modeling the activities in ICN as well [20]. In this section, a brief discussion on the use of machine learning in networking in general, and ICN in particular is carried out. ML techniques are also used to optimize the required objective (s) of a problem undertaken.

As far as the application of ML in networking is concerned, the prediction of few activities in the network including expected future traffic, types of traffic that are going to hit the network, etc. are very important from the network management point of view. Modeling of these predictions though machine learning algorithm is the best choice of current time due to advancement in machine learning techniques. The reasons are the rise in content availability and advancement in machine learning algorithms. Moreover, advances in computing capabilities is another reason for such integration of machine learning in future trend prediction. However, applying ML in network parameter prediction poses a unique set of challenges. To mention a few of them, the uniqueness of each network and lack of standard process to maintain uniformity make prediction difficult. Also, the dynamic characteristic of network applications makes prediction of traffic and other related activities difficult that is necessary for managing network operations. Adequate use of ML techniques can extract meaningful information from the historical behavior of network activities. The most promising are in networking to apply ML algorithms including traffic analysis, classification of flows, routing, and security. That is the reason the ML in networking is now an interesting area of research. In the following sections, some of the most important areas in networking to apply ML are discussed.

7.4.1 Application of ML in Communication Networks

Machine learning techniques are being applied for solving various issues in communication networks. In this section, few such applications of ML are highlighted.

7.4.1.1 Traffic Analysis and Prediction

Network traffic prediction proactively ensures the reliability of a qualitative communication network. Traffic prediction helps in many aspects, including power saving, efficient resource management in network routers, and QoS provisioning in WSN (Wireless Sensor Networks). The core Internet routers use traffic prediction to save power consumption due to computational requirements for handling heavy traffic. Higher the volume of data, more are the required numbers of processors and subsequent complexity. Such conditions in the routers result in greater power consumption. High power consumption needs an expensive cooling mechanism which in turn increases operational cost. A properly predicted network traffic helps in switching off unnecessary additional processors in these core routers during low traffic times to save power. Traffic prediction also supports network resource utilization in an efficient manner. Accurately predicted network traffic at access points is helpful to provide quality services to end users by appropriate allocation of resources. Wireless Sensor Networks also benefit from traffic prediction by saving energy. Making nodes in sleep mode in low traffic or during no activity period may significantly improve WSN lifetime. ML models are also widely used in detecting abnormal traffic patterns that may be caused by an intruder or hacker. It primarily helps in IDS or IPS by detection of DoS or DDoS attacks before it causes a severe problem to the system. Moreover, any normal congestion due to the bust user request sometimes makes network management difficult. To handle such unpredicted behavior of network traffic, ML algorithm supports network administrator up to a great extent. Such ML models assures proactive prediction with satisfactory accuracy to make the system ready for such a traffic storm. It results in lower packet loss and retransmission thereby minimizes network overhead. In absence of proper prediction, it could be detected only after the occurrence of such disaster and causing heavy packet loss and delay jitter.

The Time Series Forecasting (TSF) is the oldest traffic prediction method being used for quite a long time. This statistical model of traffic prediction can effectively forecast in the trend of expected future traffic load. This model is used to construct a regression model from the historic data in order to find a correlation between traffic volumes of past and future. As the network traffic is growing rapidly, the TSF model is becoming unsuitable and lazy to predict future traffic in current days' demand. As a consequence, people are now using machine learning approaches either independently or in combination with TSF. The most widely used ML techniques for network traffic prediction include supervised and unsupervised models. With the progress in ML techniques, various Neural Network (NN) models such as SLP (Single Layer Perceptron), MLP (Multi Layer Perceptron), RNN (Recurrent Neural Network) are found to be used in traffic prediction during the last decade. Apart from the TSF technique, traffic predictions are also be found to be accomplished using processes like frequency domain analysis of traffic flows, instead of just traffic volume. One such frequency domain analysis of network traffic flow is reported in [21]. Similarly, the work presented in [4] is one of the earliest ML-based traffic prediction methods that uses Multi Layer Perceptron (MLP)-based Neural Network (NN). In this work, the authors have shown the better accuracy of traffic prediction compared to traditional autoregressive methods. A series of mathematical analysis and subsequent proof has been given in the work in support of the proposal. The work of [22] is subsequently extended by many other researchers in [23–25] to establish the benefits of machine learning in traffic prediction. Most of these works have shown that the Single Layer Perceptron–NN acts as a universal approximator capable of approximation of continuous activity to any desired accuracy. The [26] is an MLP–NN based bandwidth prediction method proposed for grid computing environment. They have depicted the design as well as the implementation of the proposed Network Bandwidth Predictor (NBP) for a rapid and accurate network performance prediction. NBP employs a NN-based approach for network bandwidth forecasting with a design to integrate the benefits of machine learning concepts. It employs the Network Weather Service (NWS) monitoring subsystem to measure the network traffic. The NWS helps in providing an improved and accurate performance prediction with a guideline in network usage pattern. The proposed system is tested on real-time data collected by NWS monitoring subsystem and on trace files. The experimental results presented in the paper confirms that the NN-based NBP has an improved prediction compared to non-ML or AR (Augmented Reality)-based methods.

The research presented in [27] shows a Neural Network Ensemble (NNE) mechanism to predict TCP/IP traffic. The scheme uses a Time Series Forecasting (TSF) method to do such prediction. The authors have collected real-time data from two large Internet Service Providers to perform their experiments. The collected data is analyzed from different time scales and forecasting horizons were also analyzed. The proposed approach is analyzed for the viewpoint of the efficiency of several forecasting TSF approaches like Holt–Winters, the ARIMA methodology, and an NNE approach. Each of the said methods was tested under several forecasting horizons, from one to twenty-four periods ahead. A comparison among the TSF methods shows that the NNE method of prediction produces the lowest errors. However, it is necessary to consider the computational complexity into account while forecasting real-time or short-term data. In [28] an

MLP–NN-based work is found for prediction of traffic in IP network. The ANN is applied to analyze a time series of measured data for network response evaluation. Real-time Internet traffic is taken as input to identify the ANN model. Several training algorithms like gradient descent, back propagation, etc. are used to study the suitability to train then model and to estimate the weights of the neuron. The works conclude with the fact that the Levenberg–Marquardt (LM) and the Resilient Back Propagation algorithms demonstrate efficiency and accuracy of prediction. The presented results show that proposed model can be successfully used to analyze Internet traffic over IP networks. The findings of the paper help in traffic management with better accuracy.

The work discussed in [29] explores the use of Support Vector Machines (SVM) for link load forecasting. The work suggests a link load forecasting method based only on its past measurements or embedded process. They adopt a hands-on approach to evaluate SVM performance in the context of load estimation. The work depicts that the SVM is not very good in accuracy. However, by applying SVM in some modified way it could be pushed in the right direction for accuracy. In [30], a LSTM (Long Short-Term Memory)-based RNN (Recurrent Neural Network) framework to work on historical data with varying traffic pattern prop is proposed. They state that the LSTM models converge quickly and give better prediction performance. They use the LSTM model to dynamically extract features from the traffic. The idea behind using the deep learning technique to select feature automatic to make prediction accurate.

7.4.1.2 Routing

Routing is the fundamental operation inside network and extensive research efforts have been invested in broad range of terms including wide area networks, data centers, wireless networks, interdomain routing with BGP, ISP networks. In general, the route optimization deals with an unsure factor about future traffic patterns in one of the following ways: (1) Optimization of routing configurations with respect to already analyzed traffic patterns. (2) Optimization with respect to possible traffic patterns. But in the majority of cases, optimized routing configurations with respect to particular traffic patterns fails to assure good performance. Machine learning recommends to utilize the information related to past traffic patterns in order to learn the good routing configurations in future. Still the exact future traffic needs are not known in advance to the decision maker, but a realistic assumption is that the past traffic needs contain some data related to the future like traffic skewness, modification in traffic patterns across times for a day. So, a machine learning approach regularly monitor traffic needs and adapt routing strategy as per the predictions (implicit or explicit) related to future. In order to predict the future strategy, various machine learning concepts can be utilized like supervised learning approach and reinforcement learning approach. The routing of network traffic aims to choose a route for packet transmission. The selection of route will be done depending on different criteria like operation standards and purpose, such as minimization of cost, maximizes utilization of link, and provisioning of Quality of Services (QoS). The routing of traffic needs challenge abilities for machine learning models including ability to scale and cope up with the dynamic and complex topologies, the ability to learn the relationship among chosen route and the perceived value of QoS, the ability to forecast the outcomes of the decisions related to routing. The research community across the globe has investigated much to incorporate machine learning inside network traffic routing. "Reinforcement learning" is the technique of machine learning that is investigated or utilized frequently for traffic routing inside networks.

Reinforcement learning has attracted the attention of many research communities across the globe. It has many applications in the field of traffic routing in a network. The authors in [31, 32] proposed an "Q-routing" which is an application for data routing using Q-learning protocol. In this protocol, a router A learns to associate a routing procedure, like routing toward destination node X through neighbor node B, to the Q-value of it. The parameter Q is the estimated time it will take for the message to reach the node X via B, in addition with the time a packet will spend inside node A's queue plus the time needed for transmission over the route A, B. This protocol doesn't need any prior information related to network traffic pattern or topology. Though the experimental study proves that compared to shortest path first routing, it shows improvement in average message delivery time.

The authors in [33] have applied DRQ routing approach (Dual Reinforcement Q-routing) in order to minimize message delivery time. This approach integrates Q-routing with dual reinforcement learning due to which all the nodes between source and destination route get feedback in dual directions. The DRQ routing approach converges faster and this approach depicts better performance by bearing the cost of high communication overhead because of backward rewards. The authors in [34] address the issues of machine learning–driven multicast routing in mobile ad hoc networks. They have used a Q-MAP algorithm which is a Q-learning driven procedure to discover and build the multicast tree in MANETs. The exploration-free feature of this protocol leads to high routing performance because selection of actions will be done as per maximum Q-values. Still this protocol adopts a static approach which is not resilient with respect to changes in network topology.

The authors in [35] proposed a Team-Partitioned Opaque-Transition Reinforcement Learning approach (TPOT-RL). This is the first effort which applies collaborative multiagent reinforcement learning to the packet routing. Still this protocol suffers from increased computational complexity due to the exploration of a huge number of states. The overhead related to communication will also be high as each routed message will be acknowledged back by the destination node along the route from the sender for reward calculation.

7.4.2 Application of ML in Information-Centric Networks

The machine learning techniques are useful to solve many of the key issues in ICN including naming, routing, and caching. An intelligent learning mechanism in ICN helps in making decisions on caching of content or forwarding of interest packets toward the source of content. Needless to say, the integration of ML techniques in modeling the behavior of ICN is a big challenge as the ICN itself is under development from naming, routing, privacy, and security point of view. In this section of the chapter, a brief discussion is carried out on possible use of ML techniques in the areas of naming, routing, and caching. The focus is mainly to draw attention toward existing work in these areas along with a discussion of future research scopes.

7.4.2.1 ML in Caching

A multilayer perceptron (MLP) technique for analyzing entertainment traffic in a VANET (Vehicular Ad-hoc NETwork) is reported in [35]. The said deep learning–based caching mechanism is used in self-driving car to predict the traffic demand of vehicles. A Multiaccess Edge Computing (MEC) structure with deep MLP is used by the author to deploy the proposed model. The aim is to predict the types and volume of content to be accessed in a given area of operation. The authors have suggested using a roadside MEC server to log the information and user requests. The passengers' age and gender is analyzed using a CNN so that appropriate content can be delivered to the appropriate user in the car. The users request is first processed in the roadside MEC server and later classified using k-mean classifier. It enables the model to identify suitable content based on the age and gender of the passenger and to cache the same for better access. The caching problem is presented as an optimization problem with minimized content downloading delay with the application of deep learning. The Block Successive Majorization–Minimization (BS-MM) technique is applied to solve the stated problem. The proposed model is examined through simulation and results show a better accuracy of prediction in content caching.

A model called Deep-Learning-based Content Popularity Prediction (DLCPP) is proposed in [36] to cache suitable content in a content-centric architecture. The predicted popularity is later used for caching the popular contents. The attachment of DLCPP to a SDN switch makes the system distributed in nature. It also makes a reconfigurable network that employs deep learning. A metric is prepared to reflect changes in content popularity and feed to the model. The spatial–temporal probability distribution is computed by nodes in the network and fed to the auto-encoders integrated in the DLCPP. The popularity prediction is classified as a Softmax classifier and a correlation among the contents are formed. The major challenges addressed in the work include SAE structure and neuron function realization on an SDN switch. The authors have shown the DLCPP deployment on an OpenFlow SDN. The proposed caching is a lightweight caching scheme that integrates cache placement and cache replacement. The experiments presented in the paper have demonstrated a better performance of DLCPP in terms of accuracy.

A deep learning model is discussed in the context of ICN based IoT in [20]. The authors have used Convolutional Neural Network (CNN), Recurrent Neural Network (RNN), and Reinforcement Learning (RL) to observe the behavior of IoT operations. Data reliability is predicted with the help of deep learning models along with the caching accuracy. The use of deep learning in network traffic and user experience allows content caching in a real-time manner, reducing the computation time. This paper [38] develops game theoretic models to study caching possibilities between ISPs in ICN. The behavior of ISPs is analyzed in terms of noncooperative game theory. For the realization of the game theory, Nash equilibrium of the game is used. The authors claim that the said Nash equilibrium helps ISPs to learn users' access strategies and hence supports suitable caching. Through simulations it is shown that convergence to the Nash equilibrium influences the effect of caching strategies in ISPs.

The authors in [39] have proposed a neuro-fuzzy inference system-based caching scheme for CCN architecture. It utilizes centrality betweenness for selection of routers in the network and selects one with higher value as a content store. The proposed scheme is presented as a better caching mechanism which is tested across multiple network topologies. They have suggested a ubiquitous caching, considering only a selective set of routers to cache the popular content. The proposed scheme supports the caching system by finding the appropriate content router CR's location along the data delivery path. The results presented in the paper demonstrate better performance with reduced hop count and server hit ratio.

7.4.2.2   ML in Naming

A machine learning–enabled naming scheme is found in [40]. The work is focused on the naming of content produced by IoT devices, basically from camera used for weather forecasting. The CCN architecture is used for implementation of the proposed work and the hierarchical naming scheme is adopted for content identification. For implementing the naming scheme, the authors have taken 729 sentences as input weather data labels. All the sentences are then divided into words using the morphological analysis tool NLTK. They have also defined a hierarchical naming structure to name contents. The characteristics of hierarchical naming are also manipulated through ANN in work proposed in [35]. They apply MapReduce framework for content analysis and provide naming scheme for ICN architecture. The scheme provides a parallelization of content collection and analysis in the routing process through content naming. A Convolutional Neural Network (CNN) is deployed in the MapReduce architecture to analyze the ICN content for interest identification.

Although not a pure naming scheme, the work reported in [41], is an intelligent classification mechanism by means of classifying content for QoS provisioning. The proposal classifies keywords to obtain valuable information via suitable content prefixing. The intelligent function for the classification is designed using AI techniques. Various AI algorithms, including evolutionary algorithms, swarm intelligence, and ML is implemented in the work to name and maintain QoS matrices. The method is evaluated through cost function to assess their classification performances. A hybrid implementation to optimize classification is demonstrated by integration of relevant AI algorithms. The proposed method is simulated and shows that the proposed method works well in QoS performance optimization.

7.4.2.3 ML in Routing

Though ICN gives plenty of opportunities for network development of networks, there exist distinct open-ended problems in ICN that need settlement. Like the conflict among mobility and polymerizable inside naming mechanism, scalability problems related to name-based routing and an inefficient behavior of transmission control procedures with increasing number of interest messages are all worth to be addressed. Still, utilizing the analytical and computational aspects of machine learning can help the ICN to address above-mentioned issues. Still, because of adverse implications of interest and high overhead, the authors in [42] proposed the concept which utilizes the reinforcement learning approach in named data networking. They have extended and improved the existing Q-learning algorithm to solve inherent issues. They have designed and implemented two procedures, named Interest Q-Learning (IQ-Learning) and Data Q-Learning (DQ-Learning). Both these strategies will learn from past experiences and select the best routing decision. With the help of simulation study, they have proved that proposed routing strategy got higher rate for interest packet satisfaction and less delay for interest packet satisfaction. In addition to it, the proposed approach is adaptive to network topology changes compared to Best Route strategy and flooding strategy. Some of the researchers have also integrated the online machine learning procedures by adding a new element called "probabilistic binary tree structure" inside routing algorithm. The complexity of the routing procedure can be decreased and high throughput can be achieved. In addition, this approach leads to improved load balancing and less packet loss rate. In named data networking, the router will forward the interest packets for data after finding the longest prefix matches (LPM) of data names inside Forwarding Information Base (FIB). Still, the scalability of FIB with respect to large global namespace of Internet is a challenging task.

The authors in [43] proposed a novel method to interest packet forwarding that actually perform compression on FIB and convert into Artificial Neural Networks (ANNs). A bitwise tire does the splitting of namespace and assign index to ANNs. Artificial neural networks are getting trained offline by the control plane with the help of information base related to routing and matching interests. Afterwards they will be available to the data plane for the purpose of interrogation. They have also shown that the proposed approach accelerates message forwarding by specific orders of magnitude. The important point here is they have utilized artificial neural networks as processor and memory for directing messages toward next hops. So, it is actually similar to asking for directions from people about a destination. This approach is also called "ask for directions" (AFFORD). The performance of ANN–FIB is superior than the existing FIB in terms of accuracy, speed, and size of data structure. The scalability of FIB is a fundamental and major parameter in context of named data networking adoption and it is getting improved by utilizing artificial neural network concepts within it.

Looking at the importance of ML in network management, a brief introduction regarding use of ML in networking in general and in particular to ICN is presented. Various areas of networking like traffic prediction, classification, etc. are briefly explained. Along with that, a few recent researches in caching, naming, and routing for ICN are also discussed. In this section, we are not including any in-depth analysis of ML methods for ICN but brief ideas about the use of ML in ICN management have been highlighted. This section shall help researchers in finding a research direction with respect to the application of ML in ICN.

7.5   Open Research Directions for Optimization in ICN

As the problem of optimization in the context of routing, caching, and energy consumption has been an important one, there is a need to find out solutions. As the research in this line is at nascent stage, there are scopes to explore the use of various optimization techniques in the process of finding such solutions. Classical approaches such as linear programming, dynamic programming, or integer programing are being explored. Various intelligent approaches for optimization like genetic algorithm, neural network–based algorithm, or simulated annealing may also be explored. Similarly, bio-inspired optimization techniques may also be explored. Effort may also be put to detect the most suitable optimization approach to be used for a given situation of ICN. Thus, application of ML techniques for optimization in ICN is going to be a highly active research area in the days to come.

Reinforcement Learning (RL) is a subdomain of machine learning that offers a framework through which a network can learn based on its past interactions with environment to effectively choose corresponding future actions. RL has been integrated in a wide range of application domains like robotics, game playing, networks, and telecommunications, for building smart autonomous systems that improve themselves based on past experience. It is widely believed that RL is promising candidate for resolving optimization problems of distributed systems and network routing specifically. RL also incurs reasonable overhead in context of control messages, storage, and computational complexity over rest of optimization strategies for solving similar problems. There have been more than 60 protocols exist in literature since the mid-1990s with contributions in the field of optimal path selection to deliver packets in distinct types of communication networks under different users' QoS needs. RL-driven protocols aim to address features of almost all network types like MANET, VANET, IoT, WSN, CRN, NDN and DTN. The work presented in [42] explores most widely extended ML-based routing protocols (Q-routing, PQ-R learning, ARL-R, CQ-R, QoS-RSCC, etc.) in depth and related future research directions as well.

7.6 Conclusion

Optimization in protocol level is necessary for improved performance of ICN. Primarily, routing, caching, and energy consumption are the various aspects of ICN in which optimization is desired. In this chapter, the significance and importance of optimization are discussed. Some approaches already adopted for optimization of ICN are highlighted with proper references for further studies. Future research directions are also mentioned. Optimization involves modeling of the ICN system, followed by problem formulation as an optimization problem with objective function and constraints, followed by a solution approach toward the formulated problem. This is an ongoing process to find optimal solutions toward having an optimized ICN with high performance.

References

1. Pham, T.-M., Fdida, S., Antoniadis, P.: Pricing in information-centric network interconnection. In: Proceedings of the IFIP NETWORKING 2013, pp. 1–9 (May 2013)
2. Zheng, X., Wang, G., Zhao, Q.: A cache placement strategy with energy consumption optimization in information-centric networking. Futur. Internet. **11**(64), 1–16 (2019). https://doi.org/10.3390/fi11030064
3. Publish-subscribe internet routing paradigm (PSIRP) project. Website http://www.psirp.org
4. Koponen, T., Chawla, M., Chun, B.-G., Ermolinskiy, A., Kim, K.H., Shenker, S., Stoica, I.: A data-oriented (and beyond) network architecture. In: Proceedings of the ACM SIGCOMM 2007, pp. 181–192 (2007)
5. NDN project. Website http://named-data.net
6. Jacobson, V., Smetters, D.K., Thornton, J.D., Plass, M.F., Briggs, N.H., Braynard, R.L.: Networking named content. In: Proceedings of the ACM CoNEXT 2009, pp. 1–12 (Dec. 2009)
7. Li, B., Golin, M.J., Italiano, G.F., Deng, X., Sohraby, K.: On the optimal placement of web proxies in the internet. In: Proceedings of the IEEE INFOCOM 1999, vol. 3, pp. 1282–1290 (1999)
8. Kangasharju, J., Roberts, J., Ross, K.W.: Object replication strategies in content distribution networks. Comput. Commun. **25**(4), 376–383 (Mar. 2002)
9. Qiu, L., Padmanabhan, V.N., Voelker, G.M.: On the placement of web server replicas. In: Proceedings of the IEEE INFOCOM 2001, vol. 3, pp. 1587–1596 (2001)
10. Psaras, I., Clegg, R.G., Landa, R., Chai, W.K., Pavlou, G.: Modelling and evaluation of CCN-caching trees. In: Proceedings of the IFIP NETWORKING 2011, pp. 78–91. Springer (2011)
11. Pham, T.-M., Minoux, M., Fdida, S., Pilarski, M.: Optimization of Content Caching in Content-Centric Network (2017). hal-01016470v2
12. Rosensweig, E., Kurose, J., Towsley, D.: Approximate models for general cache networks. In: Proceedings of the IEEE INFOCOM 2010, pp. 1–9 (Mar. 2010)

13. Rossi, D., Rossini, G.: On sizing CCN content stores by exploiting topological information. In: Proceedings of the 2012 IEEE Conference on Computer Communications Workshops (INFOCOM WKSHPS), pp. 280–285 (Mar. 2012)
14. Araldo, A., Rossi, D., Martignon, F.: Cost-aware caching: caching more (costly items) for less (ISPS operational expenditures). IEEE Trans. Parallel Distrib. Syst. **27**(5), 1316–1330 (May 2016)
15. Wang, Y., Li, Z., Tyson, G., Uhlig, S., Xie, G.: Optimal cache allocation for content-centric networking. In: Proceedings of the ICNP 2013, pp. 1–10 (Oct. 2013)
16. Jiang, W., Ioannidis, S., Massouli'e, L., Picconi, F.: Orchestrating massively distributed CDNs. In: Proceedings of the ACM CoNEXT 2012, pp. 133–144 (2012)
17. Borst, S., Gupta, V., Walid, A.: Distributed caching algorithms for content distribution networks. In: Proceedings of the IEEE INFOCOM 2010, pp. 1478–1486 (Mar. 2010)
18. Mok, R., Chan, E., Chang, R.: Measuring the quality of experience of http video streaming. In: Proceedings of the IFIP/IEEE International Symposium on Integrated Network Management (IM), pp. 485–492 (2011)
19. Boutaba, R., et al.: A comprehensive survey on machine learning for networking: evolution, applications and research opportunities. J Internet Serv. Appl. **9**, 16 (2018)
20. Yao, H., et al.: Artificial intelligence for information-centric networks. IEEE Commun. Mag. **57**(6), 47–53 (2019)
21. Li, Y., Liu, H., Yang, W., Hu, D., Xu, W.: Inter-data-center network traffic prediction with elephant flows. In: IEEE, pp. 206–213 (2016b)
22. Yu, E., Chen, C.R.: Traffic prediction using neural networks. In: Proceedings of IEEE GLOBECOM, pp. 991–995. IEEE (1993)
23. Cybenko, G.: Approximation by superpositions of a sigmoidal function. Math. Control Signals Syst. (MCSS). **2**(4), 303–314 (1989)
24. Hornik, K.: Approximation capabilities of multilayer feedforward networks. Neural Netw. **4**(2), 251–257 (1991)
25. Funahashi, K.I.: On the approximate realization of continuous mappings by neal networks. Neural Netw. **2**(3), 183–192 (1989)
26. Eswaradass, A., Sun, X.H., Wu, M.: Network bandwidth predictor (NBP): a system for online network performance forecasting. In: Proceedings of 6th IEEE International Symposium on Cluster Computing and the Grid (CCGRID), p. 44. IEEE (2006)
27. Chabaa, S., Zeroual, A., Antari, J.: Identification and prediction of internet traffic using artificial neural networks. J. Intell. Learn. Syst. Appl. **2**(03), 147 (2010)
28. Bermolen, P., Rossi, D.: Support vector regression for link load prediction. Comput. Netw. **53**(2), 191–201 (2009)
29. Azzouni, A., Pujolle, G.: A long short-term memory recurrent neural network framework for network traffic matrix prediction. arXiv preprint arXiv:1705.05690. (2017)
30. Boyan, J.A., Littman, M.L.: Packet routing in dynamically changing networks: a reinforcement learning approach. In: Advances in Neural Information Processing Systems, pp. 671–678 (1994)
31. Littman, M., Boyan, J.: A distributed reinforcement learning scheme for network routing. In: Proceedings of the International Workshop on Applications of Neural Networks to Telecommunications, pp. 45–51. Psychology Press (1993)
32. Kumar, S., Miikkulainen, R.: Dual reinforcement q-routing: an on-line adaptive routing algorithm. In: Proceedings of the Artificial Neural Networks in Engineering Conference, pp. 231–238 (1997)
33. Sun, R., Tatsumi, S., Zhao, G.: Q-map: a novel multicast routing method in wireless ad hoc networks with multiagent reinforcement learning. In: TENCON'02. Proceedings. 2002 IEEE Region 10 Conference on Computers, Communications, Control and Power Engineering, vol. 1, pp. 667–670. IEEE (2002)
34. Ahmed, T., Coates, M., Lakhina, A.: Multivariate online anomaly detection using kernel recursive least squares. In: IEEE INFOCOM, pp. 625–633 (2007)

35. Ndikumana, A., et al.: Deep Learning Based Caching for Self-Driving Car in Multi-access Edge Computing. Networking and Internet Architecture (2018)
36. Liu, W.-X., et al.: Content Popularity Prediction and Caching for ICN: A Deep Learning Approach with SDN. IEEE Access (2017)
37. Khelifi, H., et al.: Bringing deep learning at the edge of information centric internet of things. IEEE Commun. Lett. **23**(1), 52–55 (2018)
38. Garmani, H.: Caching games between ISP in information centric network. Int. J. Control. Autom. Syst. **11**(4), 125–142 (2018)
39. Lal, N., et al.: An adaptive neuro-fuzzy inference system-based caching scheme for content-centric networking. Soft. Comput. **23**, 4459–4470 (2018)
40. Aizenberg, I.N., Aizenberg, N.N., Vanwall, J.P.: Multi-Valued and Universal Binary Neurons: Theory, Learning and Applications. Kluwer Academic Publishers, Norwell (2000)
41. Safitri, C., et al.: An intelligent content prefix classification approach for quality of service optimization in information-centric networking. Futur. Internet. **10**(4), 33 (2018)
42. Fu, B., et al.: Reinforcement learning-based algorithm for efficient and adaptive forwarding in named data networking. In: IEEE/CIC International Conference on Communications in China (ICCC), Qingdao, China, pp. 1–6 (Oct. 2017)
43. Mekinda, L., Muscariello, L.: Supervised machine learning-based routing for named data networking. In: 2016 IEEE Global Communications Conference (GLOBECOM), pp. 1–6 (Dec. 2016)

Chapter 8
A Framework for Integrating SDN in ICN

8.1 Introduction

The SDN is pushed to incorporate programmability into network routing by decoupling the control module from the data forwarding module. The control module deals with collecting routing information, and the data module deals with the packet delivery operation. Both these functionalities are accomplished through efficient programming. The said decoupling extensively supports the fast development of new routing and forwarding approaches without replacing hardware components in the core network. Integration of SDN into various network domains is gaining popularity and attention, including IoT, vehicular network, WSN, etc. Companies such as Google, CISCO, etc. are building dedicated SDN-based infrastructure to support their products for Internet services. Most of the manufacturing companies are focusing on making their products SDN-compliant for the future.

The SDN architecture adds a level of abstraction to the functionality of network switches and routers for better management. The decision-making or control mechanism is separated from the operational one to be placed on high-end computing devices rather than network switches. It certainly helps in load balancing, configuration automation, intelligent routing, and many more needs. The primary component of the control plan is an SDN controller that controls the network logic. The unified "south-bound" interface accesses the control logic for data access. The application layer comprises API to access such logic of the network control. These APIs are built upon the "north-bound" interface and offered by the SDN controller.

There any many methods adopted in NDN architecture to update the FIB table, including a fundamental flooding mechanism by an ICN router if any change is noticed. The flooding approach consumes a lot of resources, which causes network saturation and prevents scaling. Researchers are trying to adopt new methodologies to define new concepts for routing in the NDN context. The SDN is also extended to meet the routing need of the NDN. In Chap. 2, we have introduced the possible integration of SDN architecture into ICN. This chapter proposes a framework to implement the SDN routing in ICN through NDN architecture.

© Springer Nature Switzerland AG 2021
N. Dutta et al., *Information Centric Networks (ICN)*, Practical Networking,
https://doi.org/10.1007/978-3-030-46736-4_8

8.2   State-of-the-Art Work in SDN-Based ICN

In [1], a proposal for integrating SDN with ICN is presented. A novel solution for ICN and possible testbed deployments is also stated. Components of the design are segmented for feasible deployment. The first component is stated as the mapping of flow notions into an ICN. The second component in the design is concerned with realizing the core ICN functions with SDN concepts. According to the implementation stated in Click [2] router, they have also highlighted an ICN node's structure. The node's components are the Rendezvous, The Topology Manager, The Forwarding component, and The Core component. The Rendezvous component implements the respective network function. All publish/subscribe requests reach this element, match publishers with subscribers, and trigger a forwarding path. The Topology Manager handles the network topology and creates the forwarding paths. The Forwarding component implements network functions for the propagation of the content request and data chunks. According to the dissemination strategy, the Core component receives publish/subscribe requests sent by applications and other node components, forwards them to the local rendezvous component, or publishes them to the network.

In [3], a prototype of an OpenFlow-based mechanism called odICN is described. They stated various requisites for ICN in integrating SDN in the context of content-centric data center technologies. The focus of the work is to use the existing technologies to establish a deployable and manageable ICN. They argue that the network users should provide a lower delay, and the ISPs should incur a lower cost of the services and support. Therefore, ICN is supposed to fulfill the two requisites. The first one is a content identifier sent by the end users to fetch the content of their desire. And second is the network automatic resource optimization for better. The proposal is equipped with an architecture of odICN demonstrating components such as users, data centers, and OpenFlow switches in the data plane and the programmable controller in the control plane. The data plane supports users to download the expected content from the optimum data center. The controlling programs handle users' requests, locate content, and optimize deployment. The authors have presented different architectural diagrams, algorithms for operations, and message passing timing diagram to establish their claim.

Named Data Networking architecture based on Software-defined networks (NDNS) is described in [4] to inherit the benefits SDN into the NDN approach. The approach suggested separating the data plane and control plane in NDN architecture for enabling routing management by the controller. The proposal is also equipped with a cache replacement policy based on the statistical data. They have defined various architectural entities as the Global Management Information Base (GMIB), which deals with node information connected to the controller. It helps in designing a virtual topology. The Global Data Information Base (GDIB) holds the node prefixes, the Routing Information Base (RIB) contains the shortest path records between SDN switches. Another table Flow Forwarding Information Base (FlowFIB) lists the forms of actions for redirected queries. The operation model of the architecture has four steps. The first step is topology management, which collects topological information by exchanging LLDP packets between switches, nodes, and controllers. The said process basically builds the local management information base (MIB). The second step is data management, which deals with preparing and associating data prefix names for publication. The architecture publishes content prefixes without flooding. It enables the transition to NDN network scale and even data distribution. The third step is the routing that builds a local Routing Information Base (RIB) for each switch. It makes use of the Global Management Information Base (GMIB) for the RIB building. Finally, the controller builds the FlowFIB for each switch, using GDIB and RIB table that contains all name prefixes and the next interfaces to retrieve the data.

A programmable network model that supports ICN to replicate and store the files in the cloud is proposed in work [5]. The proposed architecture is composed of SDN as well as ICN layers with daemons to interconnect the OpenStack volume service (called Cinder) and the Swift Proxy. To implement the proposed model, the authors have set certain requirements. The contents of the system must be dynamically updated as no nodes have the contents of the rest of the files or cannot update volume files. The implementation of a system hides the programmable complexity of the networks for easier access to the resources. The system should be flexible and scalable enough to obtain a high level of abstraction. Moreover, the block storage should be detachable from the SDN/ICN. The architecture comprises three layers, the application, control, and the infrastructure. The Application Layer communicates with the ICN manager to establish communication with the controller. The OpenStack Neutron plugin is a daemon for creating an isolated network by tunneling hosts to storage. The control serves as an interface between the systems through an orchestrator. There exist an Infrastructure Layer that directs the routing activities based on the available information in OpenvSwitch.

The authors in [6] propose a routing strategy for information-centric networks-enabled software defined networking and community division (RISC). They have proposed a RISC system framework that contains SDN controllers, Information Centers (ICs), and CoMs. RISC uses its control plane that contains SDN controllers to take routing-related decisions and the data plane that contains ICs and CoMs to send packets. It decouples data and control planes and divides ICN topology among distinct communities. They propose a community division mechanism depending on maximal tree to fetch the data effectively. The proposed approach keeps all information related to data and forwarding inside corresponding data center for centralized control. They designed a routing strategy that contains intracommunity routing depending on same community data and intercommunity routing depending on social relation between communities. Through simulation experiments, they have proven that proposed RISC outperforms existing mechanisms like INFORM and NDNF and decreases content retrieval delay. The RISC performance is examined with respect to two realistic ICN topologies considering success times, success rate, latency, community count, stability, and throughput.

The authors in [7] propose a controller-driven routing mechanism called CRoS, which executes on top of the NDN and maintains all NDN characteristics with the help of the same interest and data messages. They have defined special identifiers and methods for controller- and router-efficient communication in NDN. CRoS defines router actions and eliminates control packet overhead by doing the coding of signaling data on content identifiers. The proposed method supports content mobility and eliminates routing information replication from controllers to content routers as they ask for routes on-demand. The protocol also needs less router memory as it stores only the paths for simultaneously consumed name prefixes. In addition to this, the mechanism automates content router provisioning and effectively installs a new path, on every content router, in a path with a one route request towards the controller. The authors have used Specification and Description Language (SDL) to describe the proposed protocol. The protocol has been validated with the concept of PetriNet, which proves that the behavior of CRoS is a deadlock or live lock free. The simulation outcomes depict that the efficiency of a protocol is robust in a case when consumer-interest rate raises with extra throughput of more communication links. The protocol efficiency is near to optimum whenever content routers operate with the required memory.

The authors in [8] have proposed an OpenFlow Compatible Key-Based Routing (OFC-KBR) mechanism inspired by a key-based routing approach in association with distributed hash tables. The KBR approach has provided data-sharing solutions for overlay networks on top of the Internet paradigm for many years now. A KBR-DHT approach is a distributed system that saves pairs of key-value, and in this any participating node can effectively fetch the value related to a given key name. The OFC-KBR protocol is a key-driven routing approach that is directly deployed at the network layer and exploits software-defined networking strength. In this protocol, endpoints are recognized with the help of virtual prefixes. These virtual prefixes are derived from a descriptive text name with no fixed format, and its format can be defined based on the needs of the service, which uses the proposed OFC-KBR protocol. The protocol is fully compatible with the existing OpenFlow standard and can work parallel with distinct L2/L3 protocols. The protocol implementation and performance evaluation are carried out by simulating realistic Internet topologies. The proposed OFC-KBR protocol's main objective is to extend the software-defined networks to non-host-centric architecture.

The proposed forwarding strategy [9] aims to solve the inherent issues that occurred by the flooding mechanism and discarding interest messages in conventional NDN. In order to do so, it exploits software-defined networking. It formulates an efficient forwarding strategy in the NDN paradigm along with distributed controllers, where decisions related to routing are taken globally. The authors have then shown the procedure for the execution of a forwarding strategy in the context of interest and data messages. They have also proposed an efficient routing mechanism that considers Quality of service (QoS)-enabled proposed forwarding method and executed within controllers. They have considered two parameters, namely network load balancing and resource consumption, and propose a genetic algorithm in order to solve the QoS constrained issue of routing with the help of information related to global network topology. A simulation study has been carried out for the demonstration of routing protocol performance. The simulation results show that the proposed

QoS routing mechanism can determine a route that has less resource consumption and good load balancing. In addition to this, the genetic algorithm performs better than the existing algorithms in the context of solving the QoS constrained routing issue.

In [10], the authors propose to exploit the Software-Defined Networking (SDN) architecture to separate the control plane and data plane and design a new routing mechanism named SRSC for CCN. The proposed solution is a clean-slate methodology that uses only CCN packets and SDN architecture. The implementation of the proposed approach is carried out inside the NS-3 simulator with an extensive set of simulation trials of the protocol proposal. SRSC exhibits superior performance over the flooding mechanism that is used by default inside CCN: it also reduces the message overhead and improves CCN performances in terms of caching. As an IP-driven solution may restrict the CCN deployment, the proposed solution is a key step toward the native CCN deployment. Apart from this, they have proposed SDN-oriented functionalities implemented directly into the CCN paradigm to facilitate the communication among controllers and nodes.

In the context of the issues related to less routing efficiency, complex control procedure, and network management in big data architecture in the conventional integrated space–terrestrial network, the authors in [11] design a satellite network architecture named software-defined information-centric satellite networking (SDICSN) depending on information-centric networking (ICN) and software-defined networking (SDN). They have designed a virtual node matrix routing mechanism (VNMR) in the SDICSN paradigm. The SDICSN paradigm experiences the flexibility of business deployment and network management by the features of the isolation of controlling and forwarding by the SDN paradigm and improves the content retrieval latency in the network through the centric of "content" as the core ICN principle. As per the predictability and periodicity of the satellite network, the VNMR mechanism gets the routing matrix by doing the relative orientation for the source and destination nodes, therefore decreasing the spatial complexity related to the input matrix for the Dijkstra algorithm and then decreasing the time complexity for the routing procedure. For forwarding information base (FIB), the procedure of combination of polling and event drove can be updated quickly in real time. Finally, the benefits of the SDICSN paradigm in routing efficiency, request aggregation, and request delay is examined through a simulation study.

8.3   The SDN Architecture for ICN

There are several SDN frameworks found for ICN implementation. In this section, the most relevant model that is closed to our proposed architecture is briefly discussed.

8.3.1   Based on OpenFlow Over CONET

The work of [12] has presented an SDN solution to support ICN. They have used the
CONET framework of ICN [13]. Their experimental design to support ICN functionality is prepared over test-bed based on OpenFlow. Following the SDN over
OpenFlow, the two primary functionalities, the caching function and the forwarding
of interest and data packets are decoupled from ICN intelligent part in the proposed
architecture. Figure 8.1 shows the architecture, as stated in [13]. The architecture
has two different planes as follows:

1. The data plane comprises of Servers, Clients, and ordinary Nodes.
2. The control plane includes a Name Routing System, controller nodes, and an
   Orchestrator node.

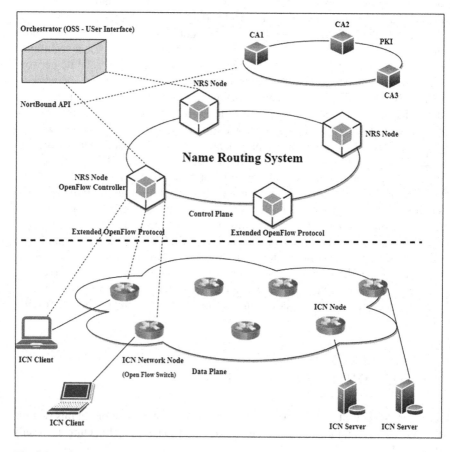

Fig. 8.1 The CORNET architecture (exclude)

An OpenFlow interface is used to communicate between the control and the data plane. Content transmission is done through the NRS nodes located in the control plane. These nodes also provide a north-bound API to the Orchestrator node to oversee the entire domain's behaving pattern. The NRS is the component of CONET, which is aligned with the SDN approach. It helps in forwarding the action of switches/routers and ensures the separation of the data forwarding plane from content routing decisions.

In this architecture, the intelligent name-based routing runs in the NRS nodes. It is implemented using OpenFlow controllers. The NRS nodes also provide an API towards the Orchestrator node. The Orchestrate controls a domain's behavior and works as a "north-bound" interface stated in SDN. The software to handle activities in ICN is executed in both servers and clients. It is patched using a CCND daemon designed by the CONET CCNx patch. This patch supports carrier packets and the transport protocol of CONET. The architecture also suggests pairing the bare switches with caching servers. The latter accomplishes the function of in-network caching and controlled through a south-bound interface. The basic caching strategy adopted in the architecture is the Tag-Based Forwarding (TBF) mechanism. In this caching policy, every content pass through an ICN-capable node is cached.

Consequently, certain limitations like lack of storage capacity, duplicate caching of same data, etc., occur in the architecture. The architecture demands a complex caching policy for effective performance considering resource constraints. The architecture facilitates deployment and tests various proposals made for SDN solutions designed for ICN architecture.

Extension of [12] is done in [14] to support ICN by exploiting SDN. They have proposed designing and implementing an open reference environment to deploy and test the ICN over SDN. They also implemented a set of caching policies to overcome the problems in [13]. These policies do not eventually demand SDN, and these could be implemented by putting the control logic in the ICN nodes. However, the use of SDN makes implementation flexible and facilitates implementation in the logically centralized SDN control infrastructure.

8.4 Proposed Architecture

In this section, an SDN framework is proposed to implement ICN. The aim of this architecture is to decouple the caching responsibilities along with the named routing. The ICN's control function is assigned to the SDN controller, and the caching of content and forwarding is assigned to the SDN switches. The proposed framework is discussed for both functional components and operational models.

8.4.1   Functional Description

The proposed SDN architecture divides the network into three areas (Fig. 8.2) :
SDN-enabled controllers, SDN-enabled switches, and ICN nodes. ICN name servers are included in the control plane, and SDN switches are considered in the data
plane. ICN nodes are actual service seekers in the topology. As we are assuming an
in-network caching, so the ICN nodes do not cache data for providing services. One
controller manages multiple switches under its coverage. We use a naming architecture as defined in [15]. The controller communicates with switches as per the specification in the communication protocol supported by OpenFlow. We have used
NDN architecture, as defined in [16]. As proposed in [17], a characteristic-based
routing algorithm is used for data packet routing. As explained in [18], a caching

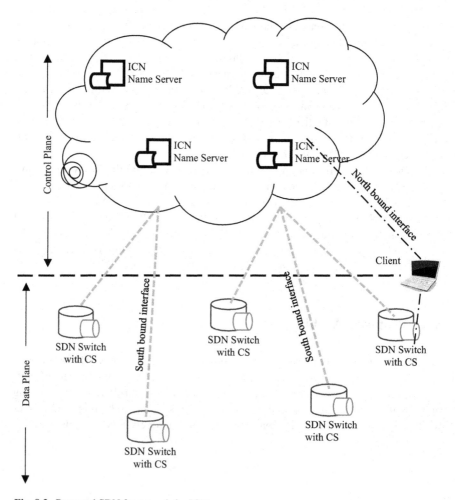

Fig. 8.2 Proposed SDN framework for ICN

mechanism is adopted for SDN provisioning in our proposed model. CRs of NDN architecture is connected to the SDN switches. It may also be integrated with the switch. Control functions include topology control, running the routing algorithm for switches, and Flow Forwarding Information Base (FlowFIB) construction for each switch. It also runs the caching-related algorithms such as caching and replacement policies for CRs in NDN architecture. The functions of CRs remain the same, as explained in our previous discussions. FIB is integrated with SDN switch rather than in CR to make the ICN SDN enabled.

Every switch in the network collects neighbor information and associated service providers in their local MIB and DIB. Such information is communicated to the control plane, and the controller builds its global information base (GIB). The controllers exchange information about their GIB to update the global topological information. From the global topological information, it fills its FIB table for necessary forwarding of interest packets. To load the FIB table, controller executes a routing table based on their name prefixes. The controller also updates the service providers name prefixes to their respective switches under its coverage. Such information is used by an SDN switch to update its FIB table.

After gathering information about neighbors and the publishers of data, the controller prepares the FIB table entry based on a name-based routing protocol. This FIB is then passed to switches under its coverage. Periodic updating of the FIB table is performed using the link–state method where the link with changed information is passed to its neighbor. The controller takes the caching decision, and it directs the SDN switch to cache the content whenever the underlying caching algorithm decides to store a particular content. To do that, the controller maintains a case storage information of each of the switches connected to it. From the table, it decides the appropriate switch for storing the content. Moreover, the computation part of the caching algorithm is carried out in the controller, making the separation of data and control part visible.

8.4.2 Routing in the Proposed Model

As described in [10], a routing model is considered in the proposed SDN model for ICN. For the implementation of the routing procedure in the proposed SDN model, the network is assumed to have N SDN-enabled routers connected to CRs. As stated above, the control plane node decides to cache content and informs the SDN switches connected to it to store the said content. It is also assumed that all the SDMN switches have enough storage to accommodate "C" pieces of content. There are total M numbers of content sources, and all are registered with any of the SDN switches in the network topology. Other than these origins of contents (i.e., the producer) and all SDN switches have cache capacity C. The size of the content is assumed to be the same, and of size U. The user produces interests with rate R. The content requests follow the independent reference model (IRM). The Dijkstra's shortest path routing as a base to forward the content requests raised by a client and implemented the proposed algorithm that runs on top of it. A client with a request for data produces an interest packet and sends it to the SDN switch attached to its network. The SDN switch checks the content in its cache table, which is managed by the connected CR component, and if available, it serves the client. Otherwise, the interest packet is forwarded to the next switch, as stated in the FIB table maintained in the SDN switch. Please note that the FIB is prepared by the SDN controller and periodically updated to the switches. Once it has found out the node from where it can download the desired content, the request gets satisfied, thereby matching the name of the content mentioned in the request packet and actual data present at the node. The corresponding node will send the related data packet to the requestor, containing the content's name and the actual content. When the data packet travels from that node to the requestor, all the controller nodes along the route run the caching algorithm to decide the need for caching that content and subsequently advised the associated SDN switch to store the data. Our motive is to retrieve the content along the path, which is the fastest possible, so that the delay can be minimized.

The SDN switch maintains a lookup table to achieve desired functionalities, storing data related to the nodes from where the requestor has recently retrieved the content. Lookup is not part of the FIB and is maintained separately from FIB. The FIB entries are a duplicate copy of the same held in the SDN controller and filled by the underlying routing protocol. The lookup table to find the location of caches and where the requested content is most likely to be present. Every tuple in the lookup table contains the content name and the node/cache's name that served the content. To preserve freshness, each row in the lookup table has a TTL value attached to it. TTL signifies the amount of time until a particular entry of the table will remain present inside the table. If the system time is higher than the TTL time for a tuple, the tuple is discarded from the table. The process of assigning TTL value is described clearly in the work of [17]. Hence the complete description of the same is skipped here. The logic for using TTL is that if some user has retrieved content recently from a particular location, there is a possibility of getting the same content again from that location. The user can route the packet toward the last found place to route the interest packet toward the origin server.

8.4.3   Caching in the Proposed Model

A popularity-based caching mechanism is used in the proposed SDN model for storing data in a content router or SDN switch. Its frequency of access determines content popularity. The universal popularity of content is computed using Zipf's law [18], taking content popularity as the basis. The Zipf's law ranks each content based on its access rate, which in turn is calculated using an exponent of popularity distribution $\propto$. The higher rank R of content implies higher access frequency and hence considered more acclaimed. The popularity of the content is directly proportional to the rank of the chunk. So, the universal popularity of the ith ranked content C_i is computed as

$$popu(c_i) = \frac{\Omega}{R_i^{\propto}}$$

(8.1)

where,

$$\Omega = \left(\sum_{i=1}^{N} \frac{1}{i^\alpha} \right)^{-1}$$

Considering N be the total contents available in the network that is being requested by different consumers. The $\propto$ determines skewness of the popularity distribution. In the original Zipf's law [18], the value of $\propto$ is considered 1; however, it is regarded as $0 < \propto < 1.5$ for broader distribution in our discussion. The popularity is directly dependent on the content request. Each content request is independently raised from the Zipf distribution. There is no correlation between any two requests for the same content or individual requests for any two different substances.

The popularity of content calculated by Eq. (8.1) represents the universal assessment. Due to the dynamism of content requests, at the time of receiving the content at the SDN controller, the data chunk's popularity may be changed. So, on receiving the data, every controller compute expected popularity to decide on content caching. The interest packet carries the popularity of the requested content as per Eq. (8.1). Whenever a controller receives an update on the popularity of content C_i, it records the same in a content popularity table. Each row in the table keeps the last m (an arbitrary count) finite (let say 10) popularities seen by the controller for the same content. Since the content request is assumed to obey Zipf's distribution and content popularity is directly influenced by request frequency, the expected popularity is modeled as a continuous random variable (X) with density function $f(x)$ represented as

$$f(x) = \lambda_i e^{-\lambda_i x}, x > 0$$

λ_i is the average access request of the content c_i. Hence, the expected popularity of the content c_i is computed as

$$E[X_i] = \int_0^T xf(x)\,dx = \int_0^T x\lambda_i e^{-\lambda_i x}\,dx$$

Considering $u = \lambda_i x$, $du = \lambda_i dx$ holds good in the equation. So,

$$E[X_i] = \frac{1}{\lambda_i} \int_0^T ue^{-u}\,du$$

(8.2)

If the expected popularity of content exceeds prespecified threshold value, it directs all the SDN switches connected to storing the content. All switches with sufficient storage will keep the content. Those with no space will execute the LRU algorithm to evacuate certain content to find a place for the new content.

8.5   Analysis of the Proposed Architecture

The architectural framework presented for ICN implementation decouples the caching and routing from forwarding decisions. The most common advantages of such separation are traffic programmability and agility. It also increases the ability to create policy-driven network supervision in terms of caching and routing policies. It also helps in allowing the creation of an ICN framework to support data-intensive applications like big data and virtualization. Naming, caching, and routing are the three major areas that need to be taken care of in ICN. The proposal of the SDN framework has described both caching and routing strategies adopted for it.

The caching mechanism suggested here is a popularity-based solution. But contents are loaded uniformly over the network. The load on CRs and the status of cache in a CR are taken into consideration for caching. Moreover, if the content is stored in the previous CR, then the current CR caches if it has ample free cache store. Otherwise, the content is not stored. The controller decides the cache of content or not. If it has to cache the content, it directs the SDN switch to cache the content. The controller maintains a case storage information of each of the switches connected to it. By performing a lookup in this table, the appropriate switch for storing the content may be selected. The caching algorithm is executed in the controller, making the separation of data and control part visible. This functionality of the proposed framework ensures the separation of control from the data plane.

The routing protocol proposed to adopt in the given framework is based on the centrality betweenness and the characteristic time. The interest packets are directed to the CR with high centrality betweenness if the lifetime of such content is not expired. In such conditions, the rate of cache hits is increased, leading to a minimized end to end delay and low traffic load for content searching.

At this point, only a theoretical presentation of the proposed SDN framework is discussed in this chapter. Although the caching and routing mechanism is tested in the simulation environments in our previous work, complete testing of the framework is left as future work. The results obtained in caching and routing are satisfactory and assume that the SDN framework will work better.

8.6   Future Research Direction in SDN-Enabled ICN

It is observed that there has been a tendency to develop Information-Centric Network (ICN) architectures based on Software-Defined Networks (SDN). Software-defined networking decouples the network architecture from infrastructure and therefore, supports continuous evolution in the network architecture in flexible manner. SDN allows a flexible control over the network infrastructure by separating the control plane from the data forwarding plane in the network devices. The core attention in ICN is the named data objects or the content of the information. Information is independent of location. ICN supports in-network caching and replication of content. However, present form of SDN does not directly support operations of ICN as it does not have any mechanism to define a fine-grained flow based on data or content identifier. There are several research issues for achieving complete realization of SDN-based ICN. Few of these research trends are mentioned below.

- The SDN switches need evolution. There is a need to have mechanisms in place to have efficient forwarding actions by the switches in the SDN–ICN ecosystem.
- SDN solution such as "openflow" should be extended to deal with the new ICN environment. While extending "openflow" keeping the compatibility with ICN issue in mind, one also needs to focus on the performance aspect of it.
- All the research challenges applicable to general form of ICN are also applicable to SDN-based ICN. The issues such as naming, routing, caching, access control, security, and privacy also need to be addressed for SDN-based ICN. While incorporating SDN and ICN to each other, obviously, the general protocols will not be directly applicable, therefore, corresponding evolutions in those are necessary.
- Incorporating ICN architecture within the framework of SDN needs converging attention toward named data object keeping the source or origin of data aside. Thus, there is a need of having mechanisms in place for data object registration and data object retrieval. Considering the application scenarios, such mechanisms may differ slightly, but overall improved performance and simplicity will be the goals of the designer.
- Reduction in the collision probability, and therefore congestion control, is an important challenge which has direct impact on the overall performance of the system. There is a need of novel congestion control protocols for SDN-based ICN.
- Ensuring Quality of Service (QoS) is another challenge. In order to achieve this there is a need to incorporate appropriate mechanisms at all layers of the protocol stack of the communication model.

As a summary, most of the research challenges mentioned above involve optimization. There is a scope to apply machine learning (ML)-based techniques in order to achieve these evolutions through SDN-based ICN. In future, it is expected that ML-based algorithms shall be developed in order to address the issues as mentioned above.

8.7 Summary

Keeping in view the impact of SDN on network appliance, in this chapter a discussion about the possibilities of integrating SDN into ICN. At the beginning of the chapter a brief overview of SDN from architecture perspective is highlighted along with the need of SDN in ICN. A prominent SDN model is discussed as well. The contribution of this chapter is the proposed framework of SDN for ICN which is presented with its architectural and functional elements. Further, a caching and routing algorithm is also proposed in the framework for the better deployment of the framework. Although, no simulation results for the work is shown, the model is justified mathematically. This framework may be a suitable for SDN-based ICN application as it separates the functionality of caching decision and actual caching in the architecture.

References

1. Syrivelis, D., et al.: Pursuing a software defined information-centric network. In: European Workshop on Software Defined Networking, Darmstadt, Germany (2012)
2. Kohler, E., et al.: The click modular router. ACM Trans. Comput. Syst. **18**(3), 263–297 (2000)
3. Chen, M., et al.: A mechanism of information-centric networking based on data centers. In: International Conference on Advanced Cloud and Big Data, Nanjing, China (2013)
4. Kalghoum, A.: Towards new information centric networking strategy based on software defined networking. In: IEEE Wireless Communications and Networking Conference, WCNC (2017)
5. Nascimento, E.B.: A programmable network architecture for information centric network using data replication in private clouds. In: 2017 IEEE 26th International Conference on Enabling Technologies: Infrastructure for Collaborative Enterprises (WETICE), Poznan, Poland, pp. 137–142 (2017)
6. Lv, J., Wang, X., Huang, M., Shi, J., Li, K., Li, J.: RISC: ICN routing mechanism incorporating SDN and community division. Comput. Netw. **123**, 88–103 (2017)
7. Torres, J.V., Alvarenga, I.D., Boutaba, R., Duarte, O.C.M.: An autonomous and efficient controller-based routing scheme for networking named-data mobility. Comput. Commun. **103**, 94–103 (May 2017)
8. de la Cruz, A., Manzanares-Lopez, P., Muñoz-Gea, J., Malgosa, J.: OpenFlow compatible key-based routing protocol: adapting SDN networks to content/service-centric paradigm. J. Netw. Syst. Manag. **27**(7), 730 (2019)
9. Li, J., Xie, R.-c., Huang, T., Sun, L.: A novel forwarding and routing mechanism design in SDN-based NDN architecture. Front. Inf. Technol. Electron. Eng. **19**(9), 1135–1150 (2018)
10. Aubry, E., Silverston, T., Chrisment, I.: SRSC: SDN-based routing scheme for CCN. In: Proceedings of the International Conference on Network Softwarization (NetSoft), pp. 1–5 (2015)
11. Liu, Z., Zhu, J., Zhang, J., Liu, Q.: Routing algorithm design of satellite network architecture based on SDN and ICN. Int. J. Satell. Commun. Netw. **38**, 1–15 (2019)
12. Salsano, S., Blefari-Melazzi, N., Detti, A., Morabito, G., Veltri, L.: Information centric networking over SDN and OpenFlow: architectural aspects and experiments on the OFELIA testbed. In: Computer Networks. Elsevier (2013)
13. Detti, A., Blefari-Melazzi, N., Salsano, S., Pomposini, M.: CONET: a content-centric internetworking architecture. In: ACM SIGCOMM Workshop ICN (2011)
14. Siracusano, G., Salsano, S., Ventre, P.L., Detti, A., Rashed, O., Blefari Melazzi, N.: A framework for experimenting ICN over SDN solutions using physical and virtual testbeds. Comput. Netw. (2018). https://doi.org/10.1016/j.comnet.2018.01.026
15. Salsano, S., Blefari Melazzi, N., Detti, A., Morabito, G., Veltri, L.: Information centric networking over SDN and OpenFlow: architectural aspects and experiments on the OFELIA testbed. Comput. Netw. **57**(16), 3207–3221 (2013)
16. Jacobson, V., Smetters, D.K., Thornton, J.D., Plass, M.F., Briggs, N.H., Braynard, R.L.: Networking named content. In: 5th International Conference on Emerging Networking Experiments and Technologies, CoNEXT, New York, NY, USA, pp. 1–12 (2009)
17. Dalasania, K., et al.: An Efficient Routing Strategy for Information Centric Networks. IEEE ANTS (2019)
18. Breslau, Lee et al. ,Web caching and zipf-like distributions: Evidence and implications. Proceedings - IEEE INFOCOM, pp. 126-134, (1999).

Chapter 9
Integrating Content Communication into Real-Life Applications

9.1 Introduction

An information-centric network is considered a novel paradigm of Internet, which demands a noteworthy contribution from the research community in this field. It provides the promising solution in order to overcome the drawbacks of existing IP-based network. Information-centric network assigns unique identifier or name to content to address the problem of insufficient address space in IP-based network. It gives access to data by its name and routes the messages by its name. It also supports caching of data at interior nodes in order to provide efficient and reliable delivery of content and self-certifying data to assure security. The well-known advantages of information-centric networks in context of efficient and speedy delivery of content and increased reliability promotes information-centric networks as a revolutionary and promising paradigm for applications in the field of Internet of things, mobile ad hoc networks, wireless sensor networks, and smart grid. Information-centric networks are an emerging solution for dissemination of data and its retrieval along with numerous benefits like good utilization of bandwidth, efficient delivery of data, and enhanced support for mobility. Information-centric network is an alternative solution to existing TCP/IP approach. In contrast to the existing IP-based network, data be considered as a first-class citizen of whole network by information-centric network. Various naming approaches are used to assign unique names to content and these identifiers can be used at the time of retrieval process.

As information-centric networks assign unique identifiers to each content chunk, the existing IPv4 addressing scheme is not sufficient. The IPv6 addressing scheme is also expected to become inappropriate in the near future. The suitability of naming scheme also depends on the base network like Internet of things, wireless sensor networks, mobile ad hoc networks, vehicular ad hoc networks, and smart grid as well as the performance requirements of the network. So existing IP-based network

© Springer Nature Switzerland AG 2021
N. Dutta et al., *Information Centric Networks (ICN)*, Practical Networking,
https://doi.org/10.1007/978-3-030-46736-4_9

paradigm does not have an efficient addressing and naming method to support a very large number of nodes with heterogeneity as well as large volume of data chunks.

During the nineteenth century, interaction with one another was noticed in variety of ways through machines, because of information communication technology services and related innovations. The major outcome of those inventions was termed "Internet," which in turn became a base for all the future scientific inventions in majority of research fields. Today, the needs of Internet users are rapidly increasing compared to the amount of efforts applied in the information and communication technology world by researchers. Like today's preferences of Skype, YouTube, FaceTime, and Facebook over traditional audio call that uses fixed connection of wires. All these above-mentioned alternative communication modes need significant bandwidth amount. It leads to an easy assumption that the faster increment in Internet usage raises certain challenges for the providers of Internet services. So, ultimately, the change is needed in base paradigm of Internet that transforms the existing host-oriented communication model to content-centric one because current Internet users are concerned about data and services and not in its providers or sources [1].

To change the focal point of communication from host to content is demand of current Internet user. Now information-centric networks can work as an alternative base architecture of Internet for range of network scenarios including mobile ad hoc networks, wireless sensor networks, vehicular ad hoc networks, smart grid, Internet of things, etc., and this requires several additional modifications or changes in existing Internet paradigm so that new Internet paradigm can emerge to satisfy the needs of users belonging to any of the above kind of network. The several recommended and major changes is as follows: communication which is energy efficient, isolation of address and identity, awareness of location, explicit cooperation for distributed services and traffic related to client–server communication, communication among person to person, integration of new framework, privacy, needs of isolation, asymmetric and symmetric Internet protocols and Quality of Service (QoS).

9.2   Recent Trends in ICN Applications

An information-centric network is an effective remedy for all the challenges and limitations of the existing TCP/IP-based Internet paradigm. Because of its various desirable features like naming, in-network caching, and name-based routing, it is most widely accepted and worked-upon paradigm of internet. Despite various advantages of this paradigm like content-centric communication, less content retrieval delay, and high data availability, there are lots of things that need to be investigated for ICN before it is used as a base paradigm with different kinds of networks like Internet of things, mobile ad hoc networks, wireless sensor networks, vehicular ad hoc networks, smart grid, etc. All these different networks have their own needs and characteristics involved with the entities present inside network which directly influence the network performance. So it is advisable to check the suitability of each aspects of ICN for any of the above networks. These aspects may cover choice of naming, caching, and routing methods in ICN-based any of the above-mentioned network scenarios, as the network performance requirements and constraints are different for different type of networks.

Information-centric network is highly prominent network paradigm for Internet of Things (IoT) scenarios due to its well-known advantages like efficient and speedy delivery of content and increased reliability. Internet of things actually connect anything and/or anyone at any given time with the help of any route on any place. Internet of things seeks attention of both research groups and industry. Internet of things is an emerging field of research and tries to connect billions of objects effectively in order to decrease involvement of people for the machine operations. To deploy information-centric networks-based Internet of things also presents several challenges like how to identify/address these resource-constrained heterogeneous objects/devices efficiently and uniquely. The following section covers an in-detail contribution of researchers who have worked in the field of ICN-based IoT applications. The ICN-based IoT approach should meet the different needs of the paradigm of the Internet of things like mobility, scalability, privacy, security, addressing, naming, interoperability, heterogeneity, energy efficiency, and availability of data [2].

Mobile ad hoc network is an independent, self-organized, and infrastructure-less multiple hop network, which comprises of battery-powered nodes in major cases. Due to this, the delivery of content among sender and receiver becomes a challenging task. As a mobile ad hoc network has topology which is dynamic in nature and has a multi-hop network, the distribution of data with reduced control messages consumes battery power and it is a much reliable approach. Numerous researches have been carried out to solve mentioned issues and still ongoing to find an efficient solution [3]. But after the evolution of new future Internet architecture called information-centric networks, the existing IP-driven, host-oriented, location-dependent network can be transformed to name-driven, content-oriented, location-independent network. And researchers have explored and investigated information-centric network-based mobile ad hoc networks. The following section covers in-detail exploration and comparative analysis for the work done in the field of ICN-based MANETs.

Vehicular ad hoc networks (VANETs) are a near-future reality, which provide significance in daily life by providing functionalities like comfort and safety of travelers during infotainment, driving, etc. The network is formed by vehicles that have capabilities related to communication. The most important characteristic related to vehicular ad hoc network is its topology, which is highly dynamic in nature. That is the reason it is a challenging job to provide an optimal and an efficient solution related to communication in vehicular ad hoc networks. Many researchers have proven the fact that TCP/IP is not an efficient protocol stack for communication inside mobile networks. And that is the reason to use wireless access in vehicular environments (WAVE) [4], a special protocol stack for vehicular ad hoc networks. With the help of WAVE short message protocol (WSMP), WAVE provides support for exchange of data excluding overhead of TCP/IP. WSMP has been designed for control packets and safety critical. In recent times, lots of research efforts have been given in order to communicate data like status of traffic, sensory data of vehicle, emergency information, and infotainment data in reliable manner and by incorporating information-centric approach which is not host oriented. The next section provides an exploration of all the recent research work done in the field of content-centric networks-based vehicular ad hoc networks, investigation of each existing approach, and present the comparative analysis among all. It also highlights the fact that how features of information-centric network can help vehicular ad hoc networks to improve the overall network performance and reduce the content retrieval delay for improvement of user-level performance parameters.

The wireless sensor networks (WSNs) are an essential component for Internet of things. A WSN comprises of a huge number of tiny battery-powered devices that have two different kinds of capabilities like communication and sensing. The wireless sensor network contains huge number of nodes/devices which are inexpensive and deployed in the areas that actually sense and keep track of physical or environmental attributes like temperature, humidity, pressure, moisture of soil, salinity of water, vital signs, etc. A single sensor device is composed of the following submodules integrated as a one module:

- Computing
- Communication
- Power source
- Sensing
- Actuation

The observed/sensed data can be sent to the collector node named sink node in a wireless ad hoc mode depending on decision-making event or at regular intervals. The collector node can be mobile or static in order to gather information from the wireless sensor network. The major goal of wireless sensor network is to keep track of the field for a longer duration of time and this can be done by saving the battery power [5]. There are different functions that need battery power like sensing, listening, processing, and wireless communication. Therefore, a sensor node must schedule its functions efficiently so that a wireless sensor node can have longer battery life. In addition to that, because of failures of nodes, the wireless sensor network architecture/topology becomes dynamic in nature. And thus a wireless sensor network should be self-organized and self-configured with respect to such node failure scenario [6]. Various solutions related to routing have been proposed in order to gain self-configuration, energy efficiency, as well as self-organization. So, the next section presents an exploration of all the recent existing research efforts and advancements in the field of information-centric network-based wireless sensor networks. A comparative analysis is also done among existing approaches in this field and how strengths of information-centric networks can make wireless sensor network operations an efficient one.

It is difficult to define the word "smart grid." It elaborates as follows: smart grid combines different modules like control and automation; advanced communication; technology that is computer driven; and systems that regulate, manage, and bring the utility electricity network which is resilient and responsive as well. The grid connects various sources of power generation and controls the need of electricity in a sustainable, economic, and reliable way. One of the major goals of smart grid is to have a balanced, need-driven electricity supply as majority of electricity generation depends on fossil fuels that actually raise the proportion of gases which are harmful inside the environment.

The smart grid is a one kind of system that has many subsystems consisting of various components. Several components have been listed and explained below [7].

- *Smart meters*: The electricity meters use the technology related to communication in order to establish connection with providers and consumers of energy to regulate need, automate billing, and find the faults for recovery on time.
- *Smart electricity production*: Various sources related to power generation from the one that use fossil fuels to renewable sources. The smart system for electricity generation produces electricity to satisfy user's need optimally, which means with least emission of carbon and related cost.
- *Smart power dissemination*: The smart power dissemination system can be done by the connection among various sources of electricity production and power consumers with the help of smart power substations and dissemination lines.
- *Smart power substations*: The smart power substations can monitor and manage noncritical and critical information related to operations like status of battery and transformer, information of breaker, security, and power factor performance.
- *Technologies related to information and communication*: These technologies give meaning for all the subsystems to communicate with one another to preserve power by effectively using, disseminating, and producing power.

Apart from these components, several other components have been depicted in below figure, and an in-detail discussion about those is outside the context of this book. In recent times, information-centric networks paradigms and their strengths have been analyzed for smart grid systems and also in order to test their effectiveness and feasibility. The next section describes each research contribution in the field of information-centric network-based smart grid. A comparative analysis of all existing approaches is shown (Fig. 9.1), including highlights of significance behind inclusion of ICN as a base paradigm for smart grid systems.

9.3 Information-Centric Networks in Internet of Things

Internet of things is composed of tiny battery-powered devices that have certain capabilities related to communication, computation, and sensing. These capabilities are associated with the objects that actually connect them with the Internet. And due to these device attachments, the IoT objects become smart to transfer and gain data to or from rest of the devices independently or with less involvement of human. In place of point-to-point communication approach, Internet of things with plenty of devices majorly emphasize on information and data. Therefore, smart IoT-based applications need contextual data that is produced either reactively or proactively

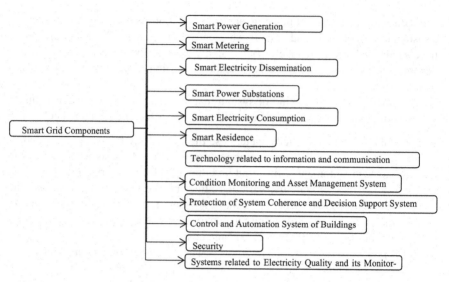

Fig. 9.1 Smart grid components

through these IoT devices [8]. These kind of networks face certain issues as mentioned below [9].

- Naming and addressing each device, dynamically
- Protocols and algorithms which are efficient in terms of energy
- Self-organization and management of network
- Interoperability norms and scalability of network
- Discovery of service and network
- Cloud connectivity and related calculation
- Dynamic merging and splitting of network
- Communication that has smaller delay value in real time
- Mass data filtering, processing, and mining
- Solutions related to security that are scalable in nature
- Majorly communication that has push-based approach
- Technologies related to privacy, security, and trust

Applications related to Internet of things have been visualized in broad range of applications as mentioned below [10].

- Intelligent healthcare system: To monitor a person anytime and from anywhere; to carry out healthcare-related tasks remotely; to sense essential signs/symptoms, monitor, control, and communicate; to give response and to perform rescue in emergency situations; to remotely diagnose and treat patient with the help of technology related to telecommunication; and to carry out smart implants of body organs
- Smart transportation: To establish communication and connection among objects like trains, vehicles, ships, tolls, traffic lights, road signs, airplanes, smart roads/ highways, etc.

- Interconnection among infrastructure and buildings as well as monitoring/tracking of the same
- Smart home
- Intelligent interconnected city
- Smart sources of power and smart grid
- Smart manufacturing process and smart products
- Sensing of data that is cooperative in nature
- Intelligent transportation and retailing of various goods.

Internet of things is creating interest among research groups because of its wide range of suitability and applicability. So the information-centric network architecture has also been explored and analyzed for Internet of things and several proposed solutions have been explored here. The authors in [11], proposed an information-centric network paradigm for Internet of things and also implemented the same. The content chunks inside IoT can be addressed with the help of unique name identifiers. The research work also highlights the advantages of implementing this approach for smart home automation system. The implementation of proposed idea utilizes push-oriented communication model for the smart home automation system.

The authors in [12] utilized the push-oriented content-centric network communication approach for traffic related to Internet of things. In the initial stage, they have discussed the scenario related to subscription. In this, subscription procedure can be executed when interest packet that contains the hierarchical identifier of the needed data can be sent without doing a PIT table entry for every interest packet forwarded. The purpose behind not doing the entry in PIT is that no data can be expected on immediate base for this interest packet. Still, the researchers have specified that creation of PIT table entry can actually avoid the problem of interest packet looping. They have used a time stamp value for each name of the content to avoid the problem of routing loops. The FIB table entries have specific field named "last_seen" that store the time stamp value. Every time when a data is forwarded, related time-stamp value can be matched with respect to this field of FIB table, and only the data that is most recent can be forwarded toward the consumers who have subscribed for it. In order to avoid the repeated forwarding of periodically sampled data, the authors have proposed a forwarding mechanism which is optimal. Inside this optimal forwarding mechanism, the data provider can pushes the data only for the said time duration that is consumer demanded. The end user can subscribe to the data that is sampled for the specific interval. All the in-between nodes can keep track for this stipulated interval with the help of counter. Data packets that have same samples and its sampling intervals or partly coinciding counters are merged and sent as a one single data packet in place of separate individual packets. At the end, they have proposed an intelligent push mechanism in order to restrict the count related to number of forwarded packets. Subscriptions having same interval use the single counter and rest of subscription durations uses individual interval counters.

The research work presented in [13] analyzes the content-centric network application for the Internet of things. They have suggested that it is not possible to implement the entire list of ICN operations on the IoT node due to their various constraints on sensing, power, memory, and processing capabilities. So, certain features like caching and security are assigned to some trusted third-party nodes in network. If the network can store only the latest sensed data received with the help of the sequence numbering or counter, then an optimal caching is achievable. As information from distinct segments of network get merged, the inclusion of sequence number for naming a data is actually not a correct solution. They have introduced new lifetime feature named "FreshnessSeconds" inside data messages to handle the above-mentioned issue. This parameter shows the amount of time till when the content can be present inside content store. They have also performed simulation of proposed ICN-based IoT scenario to check and analyze the consumption of energy. They have also examined the usage of bandwidth by varying sizes of content store and also compared the proposed approach against existing IP-based communication model.

The research work presented in [14] covers the experiments related to NDN-based IoT implementations for various corporate buildings. CCN-Lite operating system [15] is ported on the RIOT [15]. RIOT is an operating system [16] that is specifically for IoT nodes having different resource constraints. They have used hierarchical naming approach that is well suited in aggregation process of routing. They have used a simple method for flooding of interest packets called "vanilla interest flooding" ("VIF") in experiments. In this method, every node can forward the interest packet in the network whenever it gets the interest packet for the very first time. The network comprises of single producer node and multiple consumer nodes. The count related to number of radio communications for interest packets gets reduced. The FIB table, which is dynamic, can be handled by the routing approach named reactive optimistic name-driven routing (RONR). This routing method considers that the FIB table can be empty during the initialization phase in network; afterwards first interest packet gets flooded inside network and the related interest packet's name prefix can be stored inside FIB table. The following interest packets for the identical content can be autoconfigured inside FIB table and sent in a unicast way depending on the name prefix inside FIB table. They have assumed that the entire content is present inside one single provider node. But this may not be true in every situation. In order to minimize the FIB table size, there can be timeout value associated with each entry in FIB table. They have also done the comparative analysis for the effect of caching against no caching by varying number of consumers.

The research work presented in [17] proposes a new hybrid approach for naming that assigns unique names to contents with the help of flat and hierarchical components in order to provide support for both the communication model that is push and pull. This naming scheme also provides security and scalability. They have assumed smart campus scenario which is Internet of things based and initiated two different modes of transmission: (1) broadcast mode and (2) unicast mode in order to resolve the loop problem related to content-centric networks. With the help of simulation study, they have shown that proposed approach improves the interest packet transmission rate, aggregation of name, reliability, and number of traveled hops along with resolving loop problem significantly.

Majority of Internet of things proposals emphasized only on the scenario in which an interest packet can be used for the data subscription. And the content can be sent during that subscription interval. Generally in any IoT application, sensing can be performed at regular intervals, but the content may get produced as an action to any event recognized by the sensor nodes. And this can be termed as an "unrequested urgent notification." An IoT application needs this notification to be sent inside network in multicast or unicast way. There is a strong need for the information-centric network proposals that implement the support for message communication that is unrequested and urgent along with the requested content communication in Internet of things. So there are several other challenges also that need to be resolved while implementing ICN in IoT as discussed in Sect. 9.3.1.

9.3.1 Adaptation of Information-Centric Networks for Internet of Things (IoT): Challenges and Opportunities

In context of evolution of the Internet paradigm, ICN has turned up as a promising communication paradigm that is basically distinct from the existing IP address-driven paradigm. The ICN has several characteristics like data retrieval by its unique identifiers, irrespective of location of content source (IP address), and application and dissemination medium. ICN also supports in-network caching and data-based security. The ICN advantages in context of increased content distribution efficiency and security in challenging communication frameworks show the ICN potential as an innovative network architecture in the domain of IoT. The environment of IoT is challenging, majorly because of many heterogeneous and resource-constrained network devices, and unique as well as heavy congestion patterns. ICN applications in such contexts bring lots of new opportunities and need efficient design selections. In this section, a critical discussion is presented on distinct ways to achieve mentioned objectives as per the literature survey and motivation presented in previous section for ICN-based IoT.

In reality, ICN can support a variety of IoT-based applications that are content centric in nature, because ICN targets content irrespective of the object identity that originates or stores it. Let's say any traffic or environment monitoring applications are meant for particular vehicle/sensor that provides the data. The heterogeneous IoT data and services like vehicle/home services and environment information can be directly addressed by ICN identifiers. In contrast to IP addresses, these identifiers are independent of content producer's location, therefore provides content delivery by supporting mobility of nodes. ICN supports caching of content nearer to consumers, and this leads to reduction in network load and content retrieval latency. It also restricts the frequent access to resource-constrained nodes. Let's say, once the home appliance is triggered for their power consumption, the fetched data can be stored at intermediate nodes to increase availability for upcoming similar content requests.

So, the research groups are debating about deployments of ICN-IoT within the ICN research group (ICNRG) of the Internet radio task force (IRTF). Previous research such as [18, 19] are at present under debate in order to define how well-existing ICN proposals can meet the diverse IoT requirements. Meanwhile, several other research contributions were published [20–25] by keeping in mind ICN as a promising Internet paradigm in context of IoT along with the discussion about its specific feasibility aspects. The field of ICN-based IoT still needs a research attention for proper addressing of ICN-IoT integration and its related issues. This section tries to provide roadmap in order to fill this research gap. The core ICN features previously discussed have the ability to satisfy the major requirements of IoT, as summarized in Table 9.1.

9.3.1.1 Design Issues of IoT over ICN

ICN has several advantages to implement IoT applications over it. Though, to get an efficient and scalable paradigm, there are certain ICN aspects that must be taken into account while doing correct IoT design choices. ICN follows an approach of assigning names to data as well as services which are device independent in nature. This approach is desirable whenever fetching IoT content. Still, content-centric naming introduces several issues [26] that must be resolved.

1. Naming

    - *Device naming*: To name a device in IoT network is very important. The presence of actuators needs clients to act particularly on a specific device, let's say to turn it off or on. In addition to that, in order to manage and monitor devices as a part of administration procedure needs devices to have a particular identifier to uniquely recognize them. So to select an appropriate naming scheme (flat/hierarchical/hybrid) for devices is also an issue to be addressed as each naming scheme has its better suitability, merits, and demerits in context of network requirements.
    - *Size of content/service identifier*: In ICN applications, the content size is generally bigger compared to its name. In context of IoT, actuators and sensors may produce or use content which is very small, like short integer or instruction of one byte length, in order to turn on an actuator. The data name for each of these chunks of content has to recognize the content uniquely. Various existing naming mechanisms have long identifiers that have length even greater than length of actual content for different IoT applications. This can create a comparatively acceptable network overhead for larger content objects; at the same time it can become infeasible to use when size of an object is of the order of a several bytes.
    - *Hash-based data identifier*: Hashing algorithms are widely used in order to assign identifier to data so that it can be verified that the retrieved data is the one which is actually requested. This verification is possible in only cases where the needed data object exists, and there exists a directory service for identifier lookup. This mechanism is very appropriate for systems that have large volumes of content objects where data verification is critical. But this is also a challenging task for IoT-based systems in which content is generated dynamically.

Table 9.1 Major IoT needs and fundamental ICN support

Types	Named content	In-network caching	Data-oriented security	Connection-less method	Multicasting	Any casting
Quality of service: support for distinct application needs like reduced access delay		Yes			Yes	Yes
Security: Authentication, authorization, privacy, integrity, reliable	Yes		Yes			
Scalability: Avoiding the content explosion/ signaling messages when large number of devices are present	Yes	Yes		Yes	Yes	Yes
Mobility: Mobility support for requestor and data producer, support for multi-homing	Yes	Yes		Yes		Yes
Heterogeneity: Handling heterogeneity in terms of technologies or devices or procedures	Yes			Yes		
Energy efficiency: Handling communication to passive devices, including methods for lowering consumption of power.		Yes		Yes	Yes	Yes

- *Metadata-based data identifier*: Depending on metadata permits to produce an identifier for a data object before it gets created. Though this approach needs metadata-matching semantics, which is also an open issue that demands an efficient solution in context of ICN-based IoT.
- *Service naming*: Similar to device and content naming, services inside IoT networks can be assigned unique name. In contrast to HCN, this service can be facilitated inside ICN with the help of group of devices – let's say the ones that match specific metadata predicates. Another example of services contain data retrieval, get data name as input, give requested data or actuation, take an actuation instruction as input, and give a status code later on.

In IoT, the name prefix can distinguish application types, physical areas, or on the other hand other large-scale classifications that extensively distinguish gatherings of information and administrations. Still a long way from a main naming arrangement handling the referenced issues, and researchers are needed to agree upon some fundamental naming conventions.

2. Security

ICN security approaches that keep in mind the unique attributes of IoT-based applications and device restrictions strictly need to be defined. Firstly, some of the IoT-based applications need that requests from end users need to be authenticated; let's say an actuator performs an action to switch on/off any home appliances, only when it is needed by a legitimate authorized source. In recent times, ICN security methods have emphasized over data messages and do not support authentication of request messages. Secondly, IoT nodes with limited capabilities related to memory and processor rarely make use of resource-intensive procedures like public key cryptography [27].

Primary solutions against raised issues can be referred from the existing literature. As per the research work presented in [28, 29], security-related data can be adjusted in request messages in the form of last name field by utilizing the concept of virtually unrestricted hierarchical identifier composition. Still, authenticated request messages may increase the complexity related to security mechanism and increases the total identifier length, which is unsuitable for the payload capacity of resource-constrained access technologies.

In order to overcome issue mentioned below, it requires a particular lightweight solution in authentication and encryption. For this, symmetric key cryptography can be used [29]. The drawback is with the inflexibility in terms of key management, as it needs pre-dissemination of keys. Elliptic curve cryptography can become a suitable choice as it gives good trade-off for saving of resources and complexity.

In general, ICN-based security solutions specifically for IoT need to be flexible in matter of choice of cryptographic mechanisms, as there do not exist one size that fits to all solutions. The most suitable solution should be selected based on application domain and capability of device.

3. Information discovery and delivery

ICN supports identifier-based routing and lookup-based resolution systems for data discovery and this is suitable for particular IoT scenarios, majorly based on the data characteristics like popularity and dynamic generation as well as network features like infrastructured or infrastructureless [27]. NBR, integrated with content delivery, can be carried out by keeping track of some state information at every node that forwards interest packet. This approach is appropriate to access popular data in infrastructured network. It is the only feasible approach in isolated networks that have nonexistent infrastructure. It has several advantages, as mentioned below:

- Secure and resilient data retrieval with the help of adaptive forwarding integrated with in-network caching [30].
- The discovery of resource in ad hoc networks become easy via direct internode communication with the help of interest message broadcasting [31].

The drawback of NBR deployment is related to FIB explosion and increased routing updates when number of identifiers increases drastically as well as due to overhead of state information maintenance. Though identifier-prefix aggregation can resolve such issues, in integration with adaptive forwarding.

LRS can be used in infrastructure networks with popular and unpopular data with off-path caching mechanism. In real case, an intermittently grouped source IoT node can send content in a fixed always-on place like cloud to make it accessible by end users. In such scenario, to deploy a name resolution scheme depending on a hierarchically arranged DHT is favorable, like in PURSUIT and SAIL paradigms, while extra scalability features for identifier lookup can be gained with the help of capability of content center inside cloud, as directed in [32].

In conclusion, LRS and NBR approaches may complement one another. So, by utilizing multilevel hierarchical DHT, cloud computing, aggregation of identifier prefix, as well as adaptive forwarding, an efficient content discovery and delivery mechanism can be proposed with add-on scalability for large number of resources in IoT.

9.4 Information-Centric Networks in Smart Grid

In recent times, information-centric network paradigms and their strengths have been analyzed for smart-grid systems to check their effectiveness integration feasibility. In this section, all the research efforts made in this ICN based smart-grid field to automate, manage, and regulates the electricity network efficiently are briefly shown. All the existing research proposals aim to fulfill the major objective of smart-grid systems that are to provide well-balanced need-oriented supply of power, as major power generation depends on nonrenewable fuel sources.

The research work presented in [33] emphasizes on the use of information-centric networks for the applications of smart grid and recommend the use of publish–subscribe communication model in order to ease the control of smart grid with secure and simple content sharing. The smart grid system also utilizes the concept of many-to-many content-communication method among applications and devices; so, ICN is considered to be a well-suited paradigm of communication for smart-grid systems in the future. In recent times, the ICN model is adapted as an overlay for the smart-grid communication in order to support secure and seamless communication.

The authors in [34] proposed content-centric networks-oriented advanced metering infrastructure (CCN-AMI) for the smart-grid systems. To efficiently enable mobility, preserve security, and control congestion related to communication are three major goals of the proposed approach. They have also discussed the naming mechanism specifically for home network, which is an essential component of advanced metering infrastructure. The naming method contains various components like access scope (device, policy, content, and service), packet type, and mode of transmission along with data related to segmentation and version. Apart from this naming method, they have discussed several components enclosed by the content-centric network-based advanced metering infrastructure paradigm: system related to consumer energy management, data connector, smart home appliances, load management system, need supply management system, management system related to meter data, intelligent meter, etc. They have also done the performance evaluation for the CCN-AMI against existing IP-oriented intelligent-metering model. The simulation results prove that proposed CCN-based AMI approach can significantly decrease that network traffic. These researchers have also proposed a mechanism related to key management in context of CCN-based AMI in [35].

The researchers in [36] proposed and implemented the information-centric network-oriented communication model named "C-DAX," to enable content communication for smart-grid systems. They have contributed the entire C-DAX paradigm and its related components. They have also planned for a fully operational/working laboratory illustration and port the actual smart grid system's implementation in the Netherlands.

Though various research efforts have been made in the field of ICN-based smart grid as discussed above, the potential solution that can address all the challenges and meet needs of the smart-grid system as well as utilize strengths of ICN to increase throughput of overall smart grid system is yet to be proposed. Hence this field demands a quick attention from research community in order to contribute an optimized solution.

9.5   Information-Centric Networks in Wireless Sensor Networks

In recent times, the information-centric network communication paradigm has been analyzed for the applications in the field of wireless sensor networks. There have been lots of research efforts invested in this field of ICN-based WSN applications to get the optimized solution for specific application by incorporating the strengths of ICN. The researchers in [37] have proposed and implemented the information-centric network named-content communication stack with the help of operating system named "Contiki." This operating system is specifically designed for the embedded systems and wireless sensor networks that comprise of nodes/devices having various constraints on resources. The proposed approach has assumed the hierarchical naming mechanism that is identical to content-centric network. This hierarchical naming method contains name prefix and the attributes related to content as mentioned below:

Prefix: /Humidity/MEFGI
Content Attributes: /Humidity/MEFGI/WingA/Floor1/Class1
                               /Humidity/MEFGI/WingA/Floor2/Class3

The implementation of proposed approach uses two kinds of packets, interest packet and data packet, having size of 102 bytes in order to match the 127 bytes long frame for IEEE 802.15.4 standard. The processing strategy for these packets is changed in order to adjust with the processing abilities of the wireless sensor network nodes/devices. All the three data structures named FIB, CS, and PIT are also implemented by considering the processing constraints of sensor nodes in network. Performance evaluation of proposed ICN-based WSN system is done with the help of simulation study. The implementation of ICN in Contiki operating system is investigated with the help of simulations. The real-time deployment of the proposed system is evaluated using an application called synthetic monitoring for distinct sizes of network.

The research work presented in [38] covers the implementation of ICN-based communication paradigm for wireless sensor network with the help of Wiselib [39] library, which contains algorithms that generate sensor networks and that are heterogeneous in nature. ICN implementation for wireless sensor networks is called "ICN-WSN." It implements various concepts related to ICN, like interest packet, data packet, hierarchical naming, FIB, CS, and PIT. Investigation of this flexible deployment of ICN-based WSN shows how well suited ICN is for the WSN as a base architecture. The ICN paradigm for wireless sensor networks was contributed in [40]. The paradigm is decomposed into two separate tiers: The first one handles the heterogeneous devices that belong to the wireless sensor networks (sink, remote server, and sensor node). ICN is extended with certain modifications in the forwarding methods in order to improve the collection of information. The second tier is light weighted, altered, and shortened ICN forwarding method enclosing forwarding tables (PIT, FIB, and CS), packet (data and interest) transmission, and methods related to packet retransmission.

The authors in [41] discussed the application of information-centric network paradigm in wireless sensor network. To improve the reliability associated with data and interest packet communication, they have proposed the concept of defer window in context of both kinds of packets. Here the defer window for interest packet can be greater than the defer window for data packet. The delay durations for rebroadcasting of interest and data packets are I_{DW} and D_{DW}, respectively, as mentioned below:

$$D_{DW} = rand[0,DW] * DelaySlotTime$$

$$I_{DW} = (DW + rand[0,DW]) * DelaySlotTime$$

The DelaySlotTime is fixed and shorter time duration. The reason for smaller defer window of data packet than defer window of interest packet is to showcase it as a higher priority. The term defer window can also be called delay window. The basic named data networking implementation in WSN is compared against the proposed delay window driven named data networking implementation. And compared to basic NDN implementation approach for WSN, proposed approach is superior in terms of energy efficiency. The information-centric network implementation of

Contiki [42] has been deployed and simulated as well on TelosB nodes along with the discussion of results in [43]. The implementation and simulations of both the approaches confirms that proposed approach has fair content retrieval latency with lower packet exchange overhead. The researchers in [44] contributed an in detailed named data networking–driven trust model that permits both data providers and consumers to authenticate sensed content and access control methods depending on some key features and data encryption.

All the above-explained approaches focus majorly on information-centric network implementation and its feasibility in wireless sensor networks. Still, a quick and urgent attention is needed from the research community to give an optimal and an energy-efficient ICN-driven approach for wireless sensor networks.

Information-centric network approaches have many advantages for establishing communication even after an emergency or disaster has damaged the communication links [45]. The core ICN principals can provide solution to reestablish the communication in disaster situations like after an earthquake, hurricane, human-created communication breakdown, or tsunami [46, 47]. In such scenarios, communication resources and energy are at priority level because of failure of communication paths and some devices. And this situation is very critical as an efficient dissemination of disaster alert, critical rescue data to and in-between citizens should be disseminated. On a higher level, various research challenges come in context of emergency/disaster-relief situations [47, 48]. It enables the system to make use of working components of the infrastructure, even when it gets disconnected from rest entire network. It also enables usage of decentralized authentication. Data origin authentication of data fetched from the network is a challenging job when certain components of network are disconnected from any servers with security infrastructure, let's say public key infrastructure (PKI). The disaster situation also leads to several other challenging tasks like sending/receiving content across network with heavy traffic (due to any communication link or router failure), and power efficiency of router as it may need to work through batteries once a disaster has ungrouped it from power lines.

There is one recently established research project named GreenICN [49], which is funded by japan and EU. The major objective of this project is how ICN and its nodes can operate in a highly energy-efficient and scalable way. The project supports following two broader range of application scenarios: (1) To control situation after disaster when communication resources and power are at a premium concern, and it becomes critical to efficiently disseminate disaster alert and rescue data. Key point to this context is the capability to utilize fragmented networks with only irregular connectivity, the potential utilization of distinct means of communication and use of pub/sub and query/response approaches. (2) An efficient and scalable pub/sub-based delivery of video, a prime need in disaster, emergency, or normal situations.

Several ICN features (like name-based routing, authentication of named content objects, data-based access control, caching, and session-less property) make it as a prominent solution to address the above-mentioned challenges. ICN provides an efficient baseline to enable communication inside network once disaster has occurred; still it requires further investigation and research efforts.

9.6   Information-Centric Networks in Mobile Ad Hoc Networks

Mobile ad hoc network is an independent, self-organizing, and infrastructure-less network that contains majorly battery-powered nodes. These features of MANETs make the task of content delivery more complex among sender and receiver. Various research efforts have been invested to use information-centric networks as a base paradigm for MANETs because of the strong benefits of ICN architectures as discussed in Sect. 9.1.

The research work presented in [50], discuss about the new named-data networking-driven content-forwarding method in MANETs named "listen first, broadcast later" (LFBL). This method adopts three-way packet exchange policy (announcement regarding name prefixes, interest forwarding, and returning of data) that is supported by named data networking. During initial phase, a node can distributes the network-wise request packet containing the data names asked by the application. The node having the requested content can answer the query by sending corresponding data packet to the requestor. Then, receiver node can sends the positive or negative acknowledgement packet to sender. The LFBL method also stores the name prefixes that are sent inside network through request packet. Now let's say any interior node gets the request packet, it can delay its own transmission for some time duration and listen to that medium so that it can find any existing responder or forwarder node for the needed content. The simulation results prove that proposed LFBL scheme has higher content delivery efficiency compared to well-known routing protocol, which is reactive in nature, named Ad hoc on-demand distance vector" (AODV) for the standards like CSMA MAC and 802.11 MAC.

The researchers in [51] proposed and implemented the ICN paradigm for the laptop systems having Linux-based OS to setup an on-demand mobile ad hoc networks. The authors have also implemented the protocol suite which is ICN based. The proposed approach has ICN-based data structures like metadata registry, interest table, and CS. They have considered a MANET scenario for strategic as well as emergency situations, with hierarchical structure and support for group mobility. The implementation also considers hierarchical search and storage methods for the spatiotemporal and subject-based contents. During initial phase, the publisher node can distribute the same metadata content that is distributed via gateway nodes present at the upper layer in hierarchy for each individual group and stored inside registry of each node for each group. The sender can send the interest packet to gateway node. Gateway node can respond with needed data if it contains the copy. Else the interest packet can be sent toward other gateway nodes to find the data. Because of this, these gateway nodes are responsible for publishing and delivering of content.

   The authors in [52] highlighted the methods to identify basic points in the ICN-based MANET design. They have used modeling approach to evaluate the efficiency and performance of the proposed design. There are three main tasks in modeling phase: data availability announcement, send request to locate the data, and retrieve the data. They have proposed three different methods to fetch the data inside mobile ad hoc networks. The first method is reactive flooding in which the requestor node can flood out the request packet inside network to discover the data, its location, service, or service provider's location. Now, there is no case of announcement. A user just needs to flood the request packet and the data provider can respond to it with related content or service in a unicast way toward the user. The second method is proactive flooding in which each node regularly floods the data available inside MANET. The user just needs to listen for the regular announcements. Afterwards, the content request and its delivery can be completed in a unicast way. The last and third method can use the table called geographic hash tables (GHT). This approach can allocate a key for every resource inside network and when executed with the help of this key, it can give a two-dimensional coordinates pair (x,y). Firstly, the node announces (resource, host) pair to the node nearest to the key of resource. The user can first calculate the hash value for the interested resource. This can give the node location that holds the (resource, host) pair. The user interest query is then forwarded toward these nodes in a unicast way with the help of GPSR protocol. The data retrieval can be done once the resource host data is obtained. Data forwarding and retrieval are tasks that can be performed in a unicast way. The authors have also modeled data availability, delay, and cost related to overhead.

   The authors in [53] have proposed MANET in a content-centric network style (CHANET). The proposed CHANET paradigm gives in detail processing of data and interest packets having broadcast nature. It also provides various functionalities for 802.11-based mobile ad hoc network as mentioned below:

- Data decomposition and integration
- Data advertisement
- Discovery of data (interest packet) and delivery of data (data packet)
- Hierarchical naming
- Request for retransmission

   CHANET provides security and simplicity with the help of the CCN packet's broadcast nature. The nodes can listen to the broadcasted packets and delay in the time duration to decrease the collisions count. The node that obtains the packets can take the forwarding decision related to request for retransmission and interest packet. It ultimately supports the request for retransmission and method related to sequence controlling. The proposed approach also supports mobility of provider consumer.

   The research work presented in [54] covers the analysis of two different NDN-based forwarding action plans: the provider aware and provider blind. The first one is a counter-oriented broadcasting method, which delays interest and data packet transmissions to restrict the likelihood of collision. It also controls the redundancy of interest packets through overhearing the packets. The above-depicted performance is achievable for a situation where without any information about neighborhood and data source's identity, NDN packets are broadcasted inside network. In contrast to that, the second method utilizes the state information about data sources, like distance to data source, ID of data source, etc., for data retrieval process. This state information is stored inside distance table maintained at the nodes inside network. And these data can get piggybacked inside data and interest packets. All the ICN-based MANET schemes that have been explored above are summarized in Table 9.2.

9.7 Information-Centric Networks in Vehicular Ad Hoc Networks

Vehicular ad hoc network contains vehicles as network nodes that have the communication ability. This section undergoes various research efforts that have been made in the field of ICN-based VANETs.

The researchers in [55] have proposed the methods related to data decomposition–integration and reliable delivery of data. The reliability in delivery of data can be achieved by planning the interest packet retransmission for lost data packets. In order to get the recovery of lost data packets, the retransmission of interest packets can be done based on the dynamic round trip time (RTT) value for the exchange of interest packet data. The average value of RTT for a multi-hop route is predicted as a weighted moving average for round trip time samples. And depending on this data, the time out period for retransmission is planned.

The authors in [56] presented the VANET-based NDN implementation for vehicles and the simulation results for the same. Various wireless standards have been used during implementation of proposed approaches like WiMAX, Wi-Fi, IEEE 802.11p, etc. The packets can be communicated over all the existing standards to avoid the disturbance happened due to irregular connectivity. All the experiments have been conducted for a no-mobility vehicular network, vehicles roaming in a team, and the vehicles roaming around the given university campus. The authors in [57] contributed a forwarding method that retrieves information from a group of data providers with the help of digitized map data. The Navigo can bind NDN content identifiers to the geographic area of data producers. The proposed approach uses the shortest route method to forward the interest packet toward the geographic location of data provider. They have also proposed a discovery procedure for best content provider and selection method to select one among various geographic locations.

The authors in [58] proposed a traffic rules breaker ticketing application for ICN-based VANETs. Here, police officers can use the ICN interest and data packets in order to generate the violation ticket for a driver who violates the traffic rules. Several extra data structures can be used to support this application. The same researchers have proposed and examined a hash-oriented and hierarchical data

Table 9.2 ICN-based MANET schemes

Name	Method	Paradigm	Tool/simulator used	Comparison	MANET
LFBL [50]	1) Listen first, broadcast later 2) Reliable forwarding approach	Named data networking	Qualnet	AODV over CSMA-MAC and 802.11 MAC	General with random way point mobility
MANET CCN [51]	Data about data and forwarding of content in hierarchical MANETs by means of gateway nodes that control the group	Content-centric network	Implementation	OLSRD	Hierarchical paradigm and group mobility
CCM [52]	Three major tasks like announce, request, and retrieve; two types of flooding: proactive and reactive; hash table–driven approach	Content-centric network	Mathematical modeling of latency and data availability	–	General
CHANET [53]	Retransmission-enabled CCN-based MANET	Content-centric network	NS-2	FTP protocol over TCP/IP network	IEEE 802.11 standard with access point
Amadeet al. [54]	Forwarding methods like provider blind and provider aware	Named data networking	NS-3	–	Wireless network along MANET and access point

naming method for VANETs [59]. The naming method includes details like identity of data provider, distinct components that show the data attributes and the spatio-temporal resolution of data. The name also has hash value component inside it, using which data can be correctly recognized. They have also used a tri-based method for management of names so that search, add, and delete operations inside name prefix table become speedy. CCN needs name management method which can handle variable length name prefix and for this purpose, as per the performance analysis, proposed name management method is well suited.

The authors in [60] proposed robust interest forwarder selection (RUFS) method for VANETs. This method reduces the interest packet broadcast flood by choosing the next suitable forwarder vehicle node. In proposed method, every vehicle can do the sharing of statistics related to satisfied interests with neighbor nodes. These data can be managed inside neighbors satisfied list (NSL). The researchers have also summarized and discussed the challenges, problems, and related future scopes for ICN-based VANETs in [61]. All the ICN-based VANET schemes that have been explored above are summarized in Table 9.3.

Table 9.3 ICN-based VANETs schemes

Scheme name	Base ICN paradigm	Key contribution	Comparison	Type of network	Name of simulator used
CCN retransmission [55]	Content-centric network	1) Segmentation and reassembly of data 2) Data delivery in reliable manner	–	Vehicular network that is infrastructure enabled	NS-2
V-NDN [56]	Named data networking	Implementation of NDN for vehicular networks	–	Infrastructure-enabled V2V, I2V and V2I	UCLA testbed
NAVIGO [57]	Named data networking	Geographic location–based forwarding method for NDN-based vehicular ad hoc networks	GPSR	Vehicular network that is infrastructure-enabled	NS-3 (ndnSIM)
TVT [58]	Content-centric network	1) A traffic rules breaker ticketing application for ICN-based VANETs 2) Use of ICN primitives in above mentioned use case	–	Vehicular ad hoc networks	–
Hash-based and hierarchical naming [59]	Content-centric network	Hash-based and hierarchical naming method using a tri-based method for name prefix management in vehicular ICN	Bloom filters and simple trie	Vehicular ad hoc networks	C/C++
RUFS [60]	Content-centric network	Robust interest forwarder selection to reduce the interest packet broadcast flood	DR-based, CCN and NAIF	Vehicular ad hoc networks	-

9.8   Open Research in the Field of Content-Centric Real-Life Applications

Information-centric networking offers advantages in context of increased content distribution efficiency and security in challenging network scenarios. This actually claims for the ICN potential to become an effective networking architecture in the field of IoT. IoT environment is challenging because it comprises of resource constrained and heterogeneous network nodes. The integration of ICN in IoT and wireless sensor network opens door for new challenges and research directions for careful selection of design choices. The ICN-driven solution in terms of naming, security, caching, content discovery/delivery needs to be efficiently designed by taking into consideration about various IoT or sensor network requirements like scalability, quality of service, security, energy efficiency, mobility, and heterogeneity.

Information-centric mobile ad hoc networks (IC-MANETs) are a new emerging cross-cutting field of research that opens pathway for new research directions. Due to the highly dynamic network topologies and mobility of node, this network is extremely vulnerable to error or packet loss in wireless scenario. This also incurs extra overhead as routing information within FIB of IC-MANETs need to be continuously updated. The ICN-based solution for naming, security, caching, and content routing needs to be proposed by considering mobility of node and highly dynamic topology of MANET. Software-Defined Networking (SDN) is a most widely accepted paradigm that aims to change current Internet's vertical integration of control and data planes, by breaking this, isolating the control logic of network from the underlying routers and switches. This promotes logical centralization for network control. It also introduces ability for network programming. The ICN-based solution for naming, security, caching, and content-based routing needs to be formulated by considering various requirements for SDN scenarios like modularity and flexibility, interoperability and application portability, high availability, scalability, resilience, security, dependability, etc.

The ICN deployment needs to address a major challenge in poorly developed telecommunications infrastructures in which end-to-end routes are not fixed or guaranteed. In this kind of situation, delay-tolerant networks (DTN) emerged as an initial effort that effectively manages interruptions within network. In order to increase accessibility of the Internet, researchers are putting their efforts to make a universal communication architectural framework that integrates two emerging paradigms and connectivity approaches, namely DTN and ICN. The ICN-driven naming, security, caching, and content-based routing solutions need to be designed by considering various requirements for connectivity-challenged DTN world.

9.9 Summary

Because of development in early phase, information-centric networks and related paradigms, like NDN, have gone through continuous changes to make ICN applicable for different applications of each future network types. This section has covered the wide range of ICN applications in the field of Internet of things, smart grid, wireless sensor networks, mobile ad hoc networks, vehicular ad hoc networks, etc. Detailed discussions were presented on the existing and recent research work done in the above-mentioned fields by utilizing ICN as a base paradigm of Internet. The comparative analysis of the ICN-based MANETs and ICN-based VANETs schemes are also included. This literature work provides a broader vision to the readers regarding benefits and future scope of ICN in different kinds of networking applications. This section also covers an insight on how researchers have used strengths of

ICN to proposed an optimal and efficient solution in the field of IoT, smart grid, WSN, MANETs, and VANETs.

References

1. Ahmed, S.H., Bouk, S.H., Kim, D.: Content-Centric Networks An Overview, Applications And Research Challenges. Springer, Singapore (2016) ISBN: 978-981-10-0064-5
2. Miorandi, D., Sicari, S., De Pellegrini, F., Chlamtac, I.: Internet of things: vision, applications and research challenges. Ad Hoc Netw. **10**(7), 1497–1516, ISSN 1570-8705 (2012). https://doi.org/10.1016/j.adhoc.2012.02.016
3. Attia, R., Rizk, R., Ali, H.A.: Internet connectivity for mobile ad hoc network: a survey-based study. Wirel. Netw. **21**(7), 2369–2394 (2015)
4. IEEE Standard for Wireless Access in Vehicular Environments (WAVE)—Networking services—Redline. In: IEEE Std 1609.3-2010 (Revision of IEEE Std 1609.3-2007)—Redline, 30 Dec 2010, pp. 1–212
5. Singh, S.P., Sharma, S.C.: A survey on cluster-based routing protocols in wireless sensor networks. Procedia Comput. Sci. **45**, 687–695 (2015)
6. Rault, T., Bouabdallah, A., Challal, Y.: Energy efficiency in wireless sensor networks: a top-down survey. Comput. Netw. **67**(4), 104–122 (2014)
7. Katsaros, K., Chai, W., Wang, N., Pavlou, G., Bontius, H., Paolone, M.: Information-centric networking for machine-to-machine data delivery: a case study in smart grid applications. Netw. **28**(3), 58–64 (2014)
8. Miorandi, D., Sicari, S., De Pellegrini, F., Chlamtac, I.: Internet of things: vision, applications and research challenges. Ad Hoc Netw. **10**(7), 1497–1516, ISSN 1570-8705 (2012). https://doi.org/10.1016/j.adhoc.2012.02.016
9. IERC-European Research Cluster on the Internet of Things. Internet of things—Pan European Research and Innovation Vision, 2011. URL http://www.internet-of-things-research.eu/pdf/IERC_IoTPan%20European%20Research%20and%20Innovation%20Vision_2011_web.pdf (2011)
10. Wilson, S.: Rising tide—Exploring pathways to growth in the mobile semiconductor industry, 6 Nov 2013. URL http://dupress.com/articles/rising-tide-exploring-pathways-togrowth-in-the-mobile-semiconductor-industry/ (2013)
11. Waltari, O.K.: Content-centric networking in the internet of things. MSc thesis, Department of Computer Science, University of Helsinki, 25 Nov 2013. URL http://hdl.handle.net/10138/42303.
12. Francois, J., Cholez, T., Engel, T.: CCN traffic optimization for IoT. In: 2013 Fourth International Conference on the Network of the Future (NOF), 23–25 Oct 2013, pp 1–5
13. Quevedo, J., Corujo, D., Aguiar, R.: A case for ICN usage in IoT environments. In: Global Communications Conference (GLOBECOM), 2014. IEEE, 8–12 Dec 2014, pp. 2770–2775
14. Baccelli, E., Mehlis, C., Hahm, O., Schmidt, T.C., Wählisch, M.: Information centric networking in the IoT: experiments with NDN in the wild. In: Proceedings of the 1st International Conference on Information-Centric Networking (ICN'14), pp. 77–86. ACM, New York, NY, USA (2014)
15. CCN Lite: lightweight implementation of the content centric networking protocol, 2014. URL http://ccn-lite.net
16. RIOT: the friendly operating system for the internet of things. URL http://www.riot-os.org/
17. Arshad, S., Shahzaad, B., Azam, M.A., Loo, J., Ahmed, S.H., Aslam, S.: Hierarchical and flat-based hybrid naming scheme in content-centric Networks of Things. IEEE Internet Things J. **5**(2), 1070–1080 (2018)

18. Lindgren, A. et al.: Applicability and trade-offs of information-centric networking for efficient IoT. IETF Internet Draft, Jan. 2015
19. Zhang, Y. et al.: ICN based architecture for IoT—Requirements and challenges. In: IETF Internet Draft, Nov. 2014
20. Quevedo, J., Corujo, D., Aguiar, R.: Consumer driven information freshness approach for content centric networking. In: Proc. IEEE NOM (2014)
21. Amadeo, M. et al.: Named data networking for IoT: an architectural perspective. In: Proc. European Conf. Networks and Commun., Bologna, Italy (2014)
22. Baccelli, E. et al.: Information centric networking in the IoT: experiments with NDN in the wild. In: ACM Conf. Information-Centric Networking (2014)
23. Li, S. et al.: A comparative study of mobility first and NDN based ICN-IoT architectures. In: Proc. 10th IEEE Int'l. Conf. Heterogeneous Networking for Quality, Reliability, Security and Robustness, pp. 158–163 (2014)
24. Katsaros, K.V., et al.: Information-centric networking for machine-to-M chine data delivery: a case study in smart grid applications. IEEE Netw. **28**(3), 58–64 (2014)
25. Fotiou, N., Polyzos, G.C.: Realizing the Internet of Things using information-centric networking. In: Proc. 10th IEEE Int'l. Conf. Heterogeneous Networking for Quality, Reliability, Security and Robustness, pp. 193–94 (2014)
26. Lindgren, A., et al.: Design choices for the IoT in information-centric networks. In: 2016 13th IEEE Ann. Consumer Commun. & Netw. Conf. (CCNC), pp. 882–888. Las Vegas, NV (2016).
27. Amadeo, M., et al.: Information-centric networking for the Internet of Things: Challenges and opportunities. IEEE Netw. **30**(2), 92–100 (2016)
28. Burke, J. et al.: Securing instrumented environments over content-centric networking: the case of lighting control and NDN. In: Proc. IEEE NOMEN Wksp., 2013.
29. Burke, J. et al.: Secure sensing over named data networking. In: Proc. IEEE Network Computing and Applications, pp. 175–80 (2014)
30. Amadeo, M. et al.: Named data networking for IoT: an architectural perspective. In: Proc. European Conf. Networks and Commun., Bologna, Italy (2014)
31. Baccelli, E. et al.: Information centric networking in the IoT: Experiments with NDN in the Wild. In: ACM Conf. Information-Centric Networking (2014)
32. Vasilakos, X., Katsaros, K., Xylomenos, G., Cloud computing for global name-resolution in information-centric networks. In: Proc. 2nd Symp. Network Cloud Computing and Applications, 2012, IEEE, 2012, pp. 88–94
33. Katsaros, K., Chai, W., Wang, N., Pavlou, G., Bontius, H., Paolone, M.: Information-centric networking for machine-to-machine data delivery: a case study in smart grid applications. Netw. **28**(3), 58–64 (2014)
34. Yu, K., Zhu, L., Wen, Z., Mohammad, A., Zhou, Z., Sato, T.: CCN-AMI: performance evaluation of content-centric networking approach for advanced metering infrastructure in smart grid. In: 2014 IEEE International Workshop on Applied Measurements for Power Systems Proceedings (AMPS), 24–26 Sept 2014, pp. 1–6
35. Yu, K., Arifuzzaman, M., Wen, Z., Zhang, D., Sato, T.: A key management scheme for secure communications of information centric advanced metering infrastructure in smart grid. IEEE Trans. Instrum. Meas. **64**(8), 2072–2085 (2015)
36. Chai, W.K., Katsaros, K.V., Strobbe, M., Romano, P., Ge, C., Develder, C., Pavlou, G., Wang, N.: Enabling smart grid applications with ICN. In: 2nd ACM Conference on Information-Centric Networking (ICN 2015), 30 Sept–2 Oct 2015, pp. 207–208.
37. Saadallah, B., Lahmadi, A., Festor, O.: CCNx for Contiki: implementation details. Technical report RT-0432, INRIA, p. 52 (2012)
38. Ren, Z., Hail, M.A., Hellbruck, H.: CCN-WSN—A lightweight, flexible Content-Centric Networking protocol for wireless sensor networks. In: 2013 IEEE Eighth International Conference on Intelligent Sensors, Sensor Networks and Information Processing, 2–5 April 2013, pp. 123–128

39. Baumgartner, T., Chatzigiannakis, I., Fekete, S., Koninis, C., Kröller, A., Pyrgelis, A.: Wiselib: a generic algorithm library for heterogeneous sensor networks. In: Sá Silva, J., Krishnamachari, B., Boavida, F. (eds.) Proceedings of the 7th European Conference on Wireless Sensor Networks (EWSN'10), pp. 162–177. Springer, Berlin (2010)
40. Meijers, J.P., Amadeo, M., Campolo, C., Molinaro, A., Paratore, S.Y., Ruggeri, G., Booysen, M.J.: A two-tier content-centric architecture for wireless sensor networks. In: 2013 21st IEEE International Conference on Network Protocols (ICNP), 7–10 Oct 2013, pp. 1–2
41. Amadeo, M., Campolo, C., Molinaro, A., Mitton, N.: Named data networking: a natural design for data collection in wireless sensor networks. In: Wireless Days (WD), 2013 IFIP, 13–15 Nov 2013, pp. 1–6
42. Contiki: The open-source OS for the internet of things. URL http://www.contiki-os.org/
43. Abidy, Y., Saadallahy, B., Lahmadi, A., Festor, O.: Named data aggregation in wireless sensor networks. In: Network Operations and Management Symposium (NOMS), 2014 IEEE, 5–9 May 2014, pp. 1–8
44. Burke, J., Gasti, P., Nathan, N., Tsudik, G.: Secure sensing over named data networking. In: Proceedings of the 13th IEEE International Symposium on Network Computing and Applications (NCA) (2014)
45. Seedorf, J., Tagami, A., Arumaithurai, M., Koizumi, Y., Melazzi, N.B., Kutscher, D., Yagyu, T., The benefit of information centric networking for enabling communications in disaster scenarios. In: IEEE Globecom Workshops, pp. 1–7, (2015)
46. Tyson, G., et al.: Beyond content delivery: can ICNs help emergency scenarios? IEEE Netw. 28(3), 44–49 (2014)
47. Arumaithurai, M., et al.: "Using ICN in disaster scenarios," Internet Engineering Task Force, Internet-Draft draft-seedorf-icndisaster- 03, March 2015, work in progress. [Online]. Available: http://tools.ietf.org/html/draft-seedorf-icn-disaster-03
48. Seedorf, J., et al.: "Demo overview: fully decentralised authentication scheme for ICN in disaster scenarios (demonstration on mobile terminals),"in 1st ACM Conference on Information-Centric Networking (ICN-2014), (2014)
49. "Greenicn – architecture and applications of green information centric networking," website, 2014, http://greenicn.org
50. Meisel, M., Pappas, V., Zhang, L.: Ad hoc networking via named data. In: Proceedings of the fifth ACM International Workshop on Mobility in the Evolving Internet Architecture (MobiArch'10). ACM, New York, NY, USA, pp. 3–8 (2010)
51. Oh, S.Y., Lau, D., Gerla, M.: Content centric networking in tactical and emergency MANETs. In: IFIP Wireless Days (WD), 20–22 Oct 2010, pp. 1–5
52. Varvello, M., Rimac, I., Lee, U., Greenwald, L., Hilt, V.: On the design of content-centric MANETs. In: 2011 Eighth International Conference on Wireless On-Demand Network Systems and Services (WONS), 26–28 Jan 2011, pp. 1–8
53. Amadeo, M., Molinaro, A.: CHANET: a content-centric architecture for IEEE 802.11 MANETs. In: 2011 International Conference on the Network of the Future (NOF), 28–30 Nov 2011, pp. 122–127
54. Amadeo, M., Campolo, C., Molinaro, A.: Forwarding strategies in named data wireless ad hoc networks: design and evaluation. J Netw. Comput. Appl. 50, 148–158 (2015)
55. Amadeo, M., Campolo, C., Molinaro, A.: Design and analysis of a transport-level solution for content-centric VANETs. In: Proceedings of the IEEE International Conference on Communications Workshops (ICC), 9–13 June 2013, pp. 532–537
56. Grassi, G., Pesavento, D., Pau, G., Vuyyuru, R., Wakikawa, R., Zhang, L.: VANET via named data networking. In: IEEE Conference on Computer Communications Workshops (INFOCOMWKSHPS), 27 April–2 May 2014, pp. 410–415
57. Grassi, G., Pesavento, D., Pau, G., Zhang, L., Fdida, S.: Navigo: interest forwarding by geo-locations in vehicular named data networking. In: IEEE 16th International Symposium on "a World of Wireless, Mobile and Multimedia Networks" (WoWMoM), June 2015, pp. 1–10

58. Ahmed, S.H., Yaqub, M.A., Bouk, S.H., Kim, D.: Towards content-centric traffic ticketing in VANETs: an application perspective. In: 2015 Seventh international conference on ubiquitous and future networks (ICUFN), 7–10 July 2015, pp. 237–239
59. Bouk, S.H., Ahmed, S.H., Kim, D.: Hierarchical and hash-based naming with Compact Trie name management scheme for vehicular content centric networks. Comput Commun. Available online, 3 Oct 2015
60. Ahmed, S.H., Bouk, S.H., Dongkyun, K.: RUFS: RobUst forwarder selection in vehicular content-centric networks. IEEE Commun. Lett. **19**(9), 1616–1619 (2015)
61. Bouk, S.H., Ahmed, S.H., Kim, D.: Vehicular content centric network (VCCN): a survey and research challenges. In: Proceedings of the 30th annual ACM symposium on applied computing (SAC'15), pp. 695–700. ACM, New York, NY, USA (2015)